AF598674

Adaptations of Desert Organisms

Edited by J. L. Cloudsley-Thompson

Springer

Berlin
Heidelberg
New York
Barcelona
Budapest
Hong Kong
London
Milan
Paris
Santa Clara
Singapore
Tokyo

Volumes already published

Ecophysiology of the Camelidae and Desert Ruminants
By R. T. Wilson (1989)

Ecophysiology of Desert Arthropods and Reptiles
By J. L. Cloudsley-Thompson (1991)

Plant Nutrients in Desert Environments
By A. Day and K. Ludeke (1993)

Seed Germination in Desert Plants
By Y. Gutterman (1993)

Behavioural Adaptations of Desert Animals
By G. Costa (1995)

Invertebrates in Hot and Cold Arid Environments
By L. Sømme (1995)

Energetics of Desert Invertebrates
By H. Heatwole (1996)

Ecophysiology of Desert Birds
By G.L. Maclean (1996)

Plants of Desert Dunes
By A. Danin (1996)

Biotic Interactions in Arid Lands
By J.L. Cloudsley-Thompson (1996)

In preparation

Structure-Function Relations in Warm Desert Plants
By A.C. Gibson (1996)

Physiological Ecology of North American Desert Plants
By S.D. Smith, R.K. Monson, and J.E. Anderson (1996)

John L. Cloudsley-Thompson

Biotic Interactions in Arid Lands

With 38 Figures

Springer

Professor Dr. JOHN L. CLOUDSLEY-THOMPSON
Department of Biology (Darwin Building)
University College London
Gower Street
London WC1E 6BT
United Kingdom

Cover illustration: Photograph by J.L. Cloudsley-Thompson

ISBN 3-540-59261-x Springer-Verlag Berlin Heidelberg New York

Library of Congress Cataloging-in-Publication Data.
Cloudsley-Thompson, J. L.
Biotic interactions in arid lands / John L. Cloudsley-Thompson.
p. cm. -- (Adaptations of desert organisms)
Includes bibliographical references (p.) and index.
ISBN 3-540-59261-X (hardcover)
1. Arid regions ecology. 2. Biotic communities. I. Title.
II. Series.
QH541.5.D4C565 1996
574.5'2652--dc20 96-10787

Typesetting: M Masson-Scheurer, Homburg/Saar
SPIN 10125828 31/3137-5 4 3 2 1 0 - Printed on acid-free paper

In memory of P/O Hugh M. McKiernan R.A.F.

23 June 1921 - 6 August 1942

Killed during active service over the North Sea whilst piloting a Mosquito night fighter aircraft.

Acknowledgements

Warmest thanks are due to friends and colleagues throughout the world whom I have learned so much, and especially to the following, who have from read and commented upon various chapters and sections of this book: Professor Malcolm Edmunds, Professor Felicity Huntingford, Dr. Hans Kruuk, Professor Richard Tinsley and Dr. Gerald Wickens.Errors and omissions that remain are my own responsibility. In addition, I would like to thank Dr. Dieter Czeschlik and Dr. Andrea Schlitzberger, Biology Editorial Springer-Verlag, for their helpful cooperation. I am also greatly indebted to Professor Anton McLachlan and Academic Press for permission to reproduce Fig. 37, which first appeared in *Journal of Arid Environments* Vol. 21 p. 235 (1991), to Mrs Eileen Bergh for typing the manuscript, and to my dearest wife, Anne Cloudsley, without whom there would have been no pleasure in writing it.

London, Spring 1996 J. L. Cloudsley-Thompson

Contents

1 Introduction

The exigencies of life in the desert environment have resulted in the selection of a diversity of adaptations, both morphological and physiological, in the flora and fauna. At the same time, many plants and most small animals are able not merely to exist but even to thrive under desert conditions – mainly by avoiding thermal extremes and by the refinement of pre-existing abilities to economise in water. In the same way, the biotic interactions of the flora and fauna of the desert do not involve many new principles. Nevertheless, conditions in arid regions frequently do invoke refinements of the complex interrelations between predators and their prey, parasites and their hosts, as well as between herbivores and the plants upon which they feed. In this book, I shall discuss not only such interactions and their feedback effects, but also community processes and population dynamics in the desert.

The physical conditions of the desert that principally affect predators and their prey are its openness and the paucity of cover. This is restricted to scattered plants, occasional rocks, holes, and crevices in the ground. Furthermore, nightfall does not confer relative invisibility, as it does in many other ecobiomes, because of the clarity of the atmosphere. The bright starlight of the desert renders nearby objects visible even to the human eye, while an incandescent moon bathes the empty landscape with a flood of silver light. Consequently, adaptive coloration is functional at all hours of the day and night.

Many authors have commented on the high proportion of predatory species in desert faunas. This impression comes in part from the relatively high density of arachnids and reptiles, and from the fact that a large proportion of desert birds consists of carnivores. It is uncertain, however, whether this is a characteristic of all arid regions or whether it is the indirect result of the world's deserts occuring mainly at low latitudes. A considerable amount of evidence suggests an increase in the proportion of carnivorous species of animals in the fauna nearer to the equator than to the poles.

The scope of this book is large, and I have therefore found it expedient to illustrate many of the topics discussed by the use of selected examples only; and because so many taxa are involved, I have cited the families of most genera and species when first mentioned. As fas as possible, attention has been drawn to key works and reviews that provide a convenient entrée to the appropriate literature on each subject. Even so, the bibliography is quite extensive, and any attempt at complete coverage would have rendered the volume unnecessarily unwieldly. Critics will, no doubt, be happy to draw attention to numerous important points that should not have been omitted; this will be helpful. At the same time, there is inevitably some overlap with other books in this Series, notably those of G. Costa (1995) and Heatwole (1996). This is not as detrimental as it may at first appear, however, because our approaches to the same topics are usually rather different and I have attempted to supplement rather than repeat their data. Moreover, Costa has concentrated on intraspecific relationships, while I have paid more attention to interspecific relations.

Some species are endemic to arid regions, but others, such as eland, gemsbok, springbok, rhino, elephant and lion, range throughout more mesic biomes but are to be found in the driest of deserts as well. This is also true of a number of species of birds which move into and out of the desert in response to seasonal or unpredictable abundance of food. Some of the work on such species has been cited in the following pages, even though it was carried out in less arid parts of the species' ranges. At the samc time, aspects of topics already covered by, or due to be discussed in other volumes of this Series receive proportionally less attention.

Throughout the book, I have referred to adaptation and selection within a neo-Darwinian context, in which competition has provided one of the major forces for change. Although this traditional view of the role of competition in macro-evolutionary theory has been challenged within recent years, my own views in this respect are conformist and conventional. It is, perhaps, better to be old-fashioned and correct than to follow new trends that could well lead in the wrong direction, merely because their novelty appeals to the imagination. Despite the very large amount of information now available, there are still numerous points of disagreement among biologists and enormous gaps in knowledge. The present book has a somewhat personal bias.

1.1 Evolutionary Parallels

More homogeneous environments normally contain fewer species of plants and animals than do less homogeneous environments. Moreover, the most numerous species of animals tend to occupy extensive habitats, such as wide expanses of grassland or, in the desert ecobiome, a sea of sand dunes rather than, say, a single tuft of perennial grass. The diversity of desert organisms, therefore, lies not so much in diversity of form and number of species as in the frequency with which a particular shape or physiological adaptation has been evolved independently – in different taxonomic groups that have responded to almost identical environmental conditions in different parts of the world yet in remarkably similar ways (Cloudsley-Thompson 1993a).

The extreme physical and climatic conditions of the desert biome have engendered or enhanced a number of interrelated morphological, behavioural, and physiological adaptations, many of which are paralleled in various quite unrelated taxa of animals and even of plants. For example, the integuments of desert arthropods and the cuticles of desert angiosperms both possess extremely impervious wax layers, while insect spiracles and plant stomata are especially well developed among desert species. There is also a positive correlation between the high critical temperatures of arthropod cuticles and the high ambient temperatures that the animals experience in their natural habitats.

Xeromorphic desert plants are characterized by having their stomata buried in the epicuticle below the general level of the plant surface, or superficially depressed due to the extremley thick surrounding cuticle, while the spiracles of desert arthropods are similary often sunken or hidden below the surface of the integument (Hadley 1972). In xerophilous buprestid beetles, the spiracular openings are covered by a basket-work of cuticular outgrowths which are believed to impede the diffusion of water molecules to a greater extent than those of carbon dioxide. The very fine hairs and spines of some cacti may serve a similiar function. Moreover, they reduce heating by reflecting incident radiation and creating a thick boundary layer (Louw and Seely 1982).

Again, to give other examples of parallelism between diverse organisms: the leaves and stems of desert plants are frequently oriented so that the heat load from solar radiation is reduced. There is a parallel here with the orientation of the "compass" nests of termites (*Amitermes* spp.) in the Australian desert. These are oriented in a north-south direction, as are the pads and stems of cacti. This results in their being warmed when the ambient temperature is low while, at midday, a relatively small area faces the sun. Steep leaf angles also result in reduced solar heating (Gibson 1996).

Locusts orient their bodies at right angles to the sun's rays in the morning and evening. At midday, however, when the heat is great, they turn their heads towards the sun, thereby reducing the surface exposed to radiation. Ostriches, camels, wildebeest, springbok and probably many other species of antelope do the same (Louw and Seely 1982; Hofmeyr and Louw 1987): so do reptiles; while the leaves of plants, through changes in turgor, can reduce the effective area so that excessive transpiration and wilting are avoided (Cloudsley-Thompson 1994).

Many other examples of parallelism between plants and animals could be cited. Before taking flight, hawkmoths (Sphingidae) raise the temperature of their thorax by 20 °C above the ambient by rapidly vibrating their wings and generating heat metabolically. This ensures optimum flight efficiency (Heinrich 1993). The spadix of the mesic voodoo lily (*Sauromatum venosum*: Araceae) may also, metabolically, reach up to 15 °C above the ambient temperature. This volatilizes the amines that are responsible for the horrible smell which attracts the flies that pollinate the flowers (see Sect. 7.4.4). When they are incubating their eggs, female Indian pythons (*Python molurus*) and other Boidae warm the clutch by physiological thermoregulation, rapidly flexing their body muscles when ambient temperatures fall below 33 °C (see Cloudsley-Thompson 1971, 1991, 1993a). Again, termite nests are warmed by metabolic heat produced in the fungal gardens – one could go on almost indefinitely citing pertinent instances of metabolic heat generation among organisms other than homoiothermic birds and mammals. What is possibly one of the most surprising examples of convergence is afforded by the parallel between termite colonies, and colonies of naked mole rats (*Heterocephalus glaber*: Bathyergidae), whose behaviour resembles that of social insects even to the extent that reproduction is confined to a "queen" mole rat, the only breeding female in the entire colony (Sherman et al. 1991; Sect. 7.1.1).

1.2 Ecological Analogues

Different taxa often respond to environmental factors in remarkably similar ways. When quite unrelated species come to look alike as a result of parallel evolution in different geographical regions, they may be known as "ecological equivalents" or more accurately, as "ecological analogues". One of the best-known examples is the fennec fox (*Fennecus zerda*) of the Sahara, which has a number of adaptive characteristics that parallel those of the American kit foxes (*Vulpes velox* and *V. macrotis*). Similarly, there is a close structural similarity between the kangaroo rats (*Dipodomys* spp.) of North America and the Old World jerboas (*Jaculus* and *Dipus* spp.). Reptilian examples include the Australian thorny devil *Moloch horridus* (Agamidae) and its ecological analogue *Phrynosoma platyrhinos* (Iguanidae), the horned lizard of the North American deserts, which likewise exploits a diet of ants. These are anatomically closer to one another than either is to any other member of its own lizard taxon. No lizard of any other desert region of the world has adopted a similar life style (Pianka 1985, 1986).

Another well-known reptilian example of ecological analogues is provided by the North American sidewinder (*Crotalus cerastes*: Crotalinae) and its Palaearctic counterpart, the Saharan sand viper (*Cerastes cerastes*: Viperinae). Both these snakes move by throwing lateral loops forward, and hide themselves by flattening their bodies and shovelling sand over their backs. They are so much alike in general appearance that, were it not for the rattle of *Crotalus cerastes* and the pits between its eyes and nostrils, the two might easily be confused (Cloudsley-Thompson 1991). The Namib desert sidewinder *Bitis peringueyi* is also similar in appearance (Louw and Seely 1982).

In his analysis of the ecological niche and community structure of the lizard faunas of the North American, Kalahari and Australian deserts, Pianka (1985, 1986) pointed out that both North America and Australia have long-legged species that frequent the open spaces between plants – the iguanid *Callisaurus draconoides* and *Amphibolurus* (= *Ctenophorus*) spp. (Agamidae) respectively – while each region has a medium-sized lizard-eating species, *Crotaphytus wislizeni* (Iguanidae) in North America, and *Varanus eremius* (Varanidae) in Australia. A few Kalahari-Australia species pairs are also roughly equivalent, e.g. the subterranean skinks *Typhlosaurus* and *Lerista* spp. (Scincidae) and the semi-arboreal *Agama hispida* and *Pogona minor* (Agamidae).

In addition to the examples already mentioned, other members of the Agamidae frequently occupy ecological niches similar or analogous to those of Iguanidae. Examples include *Uromastyx* spp. (Agamidae) of the Great Palaearctic desert and North American *Sauromalus* and *Cachryx* spp. (Iguanidae). The south-east Asian agamids *Leiolepis* spp. resemble North American iguanids such as *Dipsosaurus dorsalis* and the Australian *Amphibolurus pictus* (Agamidae), while the Iranian agamid *Phrynocephalus mystaceus* and the African gecko *Geckonia chalaziae* may also have many characters in common. Despite this, Pianka (1985, 1986) concluded that the differences between the ecologies of most lizard species in the three continental deserts that he studied are much more striking than the similarities. "It is easy to make too much out of convergence, and one must always be wary of imposing it upon the system under consideration".

Nevertheless, recognition of convergence is an important factor in the understanding of natural selection. Another example of convergence is provided by the adaptive coloration of desert animals, including reptiles, which almost always match the sandy hues of their environment (Sect. 3.4) or else are black (Sect. 3.8; Cloudsley-Thompson 1979). It is probably a tautology to claim that a particular ecobiome, such as the desert, should engender comparable adaptations in its fauna in different zoogeographical realms of the world. Pianka (1986), however, provides an excellent example, based on scorpion predation. While scorpions are solitary prey items, they are extremely large and nutritious, thereby presumably facilitating the evolution of dietary specialization. In the Kalahari they are preyed on by *Nucras tessellata* (Lacertidae) and in Australia by *Pygopus nigriceps* (Pygopodidae). The diurnal *N. tessellata* forages widely to capture these animals in their daytime retreats, whereas the nocturnal *P. nigriceps* sits and waits for scorpions moving at night, above ground, during their normal period of activity. No North American desert lizard specializes on a diet of scorpions, but the small snake *Chionactis occipitalis* (Colubridae) appears to have usurped this particular ecological role.

Whereas some lizards have evolved as dietary specialists, rather more are generalists. *Moloch horridus* and *Phrynosoma* spp. eat essentially nothing but ants. The Kalahari lizards *Mesalina lugubris* (Lacertidae) and *Typhlosaurus* spp. (Scincidae) feed entirely on termites, as do the Australian nocturnal geckoes *Diplodactylus conspicillatus* and *Rhynchoedura* spp., as well as some day-active *Ctenotus* spp. (Scincidae; Pianke 1986). Dietary diversity occurs in many species of lizards which eat almost everything they can catch and overcome. Variations in diet also occur within the same species, both from time to time and from place to place, as opportunities present themselves and the abundance of particular prey species fluctuates.

Convergence is less marked among arthropods than among reptiles, probably because the former are smaller, more diverse, and therefore occupy less generalized ecological niches. Nevertheless, when comparisons are made between the scorpion fauna of the Kalahari and Namib deserts and that of the northern Sahara, it is found that, in contrast to the preponderant species richness of seven genera of Scorpionidae in southern Africa, there is perhaps only one species in the north, namely *Scorpio maurus*. This however, is either polymorphic or else is represented by numerous subspecies, each with differing life-styles, depending upon the texture of the soil (Cloudsley-Thompson and Lourenço 1994). These variations or subspecies are comparable with analogous species of the genus *Opisthophthalmus* in southern Africa. At the same time, the dune-inhabiting *O. flavescens* of the Namib occupies a niche comparable with that of the psammophilous *Buthacus arenicola* of the Sahara (Cloudsley-Thompson 1991).

Remarkable morphological similarities are apparent among all the species of scorpions, as among those of Pseudoscorpiones, Thelyphonida, Amblypygi, Palpigradi and Schizomida. This renders the detection of convergence rather difficult. There are greater differences, however, between the various families of Solifugae. The short legs of the Rhagodidae and of most Hexisopodidae, for instance, contrast with the longer legs found in other families (Kaestner 1968). With regard to harvester ants, Medel (1995) concluded that constraints, related to the evolutionary history of each species assemblage, inhibited convergent evolution in response to local selective pressures.

There are more desert-dwelling species of mammals in North America (109) than in Australia (73). The difference is not due to the comparative ages, sizes or physiographics of the two desert regions. According to Morton (1979), it is a result of the abundance of graminivorous species in North America. Graminivorous ants and birds are more diverse in the Australian deserts and have probably secured their grain-eating role at the expense of mammalian granivores. In contrast, insectivorous mammals are more common in Australia: in North America their role is at least partly filled by omnivorous rodents. The difference between the two faunas suggests that convergence between the two arid regions is limited. As far as birds are concerned, North America has four carnivorous species compared with 27 in Australia; 31 granivorous/insectivorous species compared with 25; and 30 ground-feeding insectivorous species compared with 40 in Australia. Morton (1979) also estimated that the numbers of lizard species are 57 to 143 and of ants 161 to 300–400 in the two desert regions respectively.

Ecological analogues are not confined to the animal kingdom. There are numerous instances in which Cactaceae of the New World are strikingly similiar in appearance and mode of life to Asclepiadaceae, Euphorbiaceae, Chenopodiaceae and other succulent plants of the Old World. Species with ribbed, columnar or globose stems bearing leaves or stems modified into spines or thorns can be found in both cacti and Euphorbiaceae, presumably in response to similiar selective pressures for water storage, for reduction of surface area, as adaptations for convectional dissipation of heat, and as protection against herbivores (Niklas 1992). Cacti characterize the American deserts, but large succulents are less common in the Great Palaearctic desert because the metabolic work necessary to give rise to larger desert plants cannot be undertaken under extremely arid conditions. Nevertheless, *Euphorbia echinus* and *Caralluma* spp. (Danin 1996), for instance, closely resemble cacti and are clearly convergent forms (Grenot 1974). Likewise, *Euphorbia abyssinica* (Euphorbiaceae) of the Red Sea hills could well be an ecological analogue of the saguaro (*Carnegiea gigantea*) (Cactaceae; Fig. 7) and other columnar cacti of Arizona. Protective resemblance and mimicry in plants are discussed below (Sect. 6.1).

Such parallels as these are not confined to morphological and physiological similarities between different species. They are also apparent in the interactions between different species, whether of plants or animals. They can be found within the realms of predation, parasitism, herbivory, and in the community processes outlined in the following chapters. It is important, nevertheless, to distinguish between analogies and identities, however useful the former concept may be. Even Charles Darwin's theory of natural selection depends to a considerable extent upon the analogy between the controlled breeding of domesticated plants and animals and the historical development of the organic world. Failure to recognise this is a weakness of his theory.

Compared with the exuberance and diversity of colour and form among the plants and animals of the tropical rainforest, the sombre hues and structural parallelism evident among desert organisms must appear to the casual observer to be remarkably dull. Nevertheless, as I have attempted to indicate in the introductory chapter to *Sahara Desert*

(Cloudsley-Thompson 1984a), arid environments are places of wonder and beauty in their own right. The symmetry imposed by sonata form in its first movement does not render a great symphony any less beautiful!

2 Predatory Techniques

Predators must catch and kill their prey before they can eat it. They do this in different ways, using a wide variety of weapons – the scorpion's sting, the spider's web, the sidewinder's aggressive crypsis and venom, or the flight and stoop of the falcon. Successful predation involves at least five stages: detection of the prey as an object distinct from its background; identification, despite crypsis (Sect. 3.4), aposematism (Sect. 3.8), disguise (Sect. 3.5), or mimicry (Sect. 3.6); approach despite flight (Sect. 4.1), deimatic displays (Sect. 4.7) and so on; subjugation, despite armour (Sect. 3.7), weapons (Sect. 4.6) and other secondary antipredator defences, including autotomy (Sect. 4.4) and thanatosis (Sect. 4.2); and, finally, its consumption (Endler 1986). Animals are almost invariably endowed with multiple defence mechanisms. Some predators seek their prey actively by searching for it (Sect. 2.1); others lie in wait, occasionally in disguise, and ambush it (Sect. 2.2). Finally, certain predators, including vultures, scarcely catch living prey at all, but have secondarily adopted a scavenging mode of life (Sect. 2.3). Combinations of predatory technique are frequently employed.

Natural selection, generated by competition between predators for food and their prey to avoid being eaten, leads to an "arms race" between predator and prey (Cloudsley-Thompson 1980; J. Owen 1980). All animals, even top predators, at some stage of their life cycle are potential prey for some other predator. Consequently, their adaptations represent a compromise between being effective in hunting and in escaping capture. For example, mantids are raptatorial predators as far as other insects are concerned, yet are themselves the prey of insectivorous birds. The predators of invertebrates have been discussed by Heatwole (1996). Those of insects, spiders and other small arthropods fall into two categories: small (other arthropods) and large (vertebrates), against each of which different types of defence have been evolved (Chap. 3, 4). At the same time, an animal's role as a predator is modified by its own potential as prey. In general, predators have greater numerical effects when prey populations are low than when they are higher, while predation affects the structure of prey populations whose survival is longer when predators are excluded (Meserve et al. 1993).

2.1 Active Searching for and Stalking Prey

The ethology of predation has been analysed in considerable depth by Curio (1976), and no attempt will be made here to elaborate on this aspect of the subject. Many predators hunt by speculation, moving apparently at random until they see, smell, hear, touch, or, in some other way, detect their prey. Even large speculatory hunters often increase their chances of finding prey by searching in especially promising places, such as water holes or patches of succulent vegetation. Raptors tend to hunt in specific areas that are frequented by potential prey. Furthermore, erratic or protean behaviour, especially in open country, on the part of predators such as foxes and jackals when hunting small rodents, may help by confusing they prey (Driver and Humphries 1988). Active searching for prey occurs in all taxa of predators in the desert ecobiome and depends upon locomotion. Searching behaviour has been reviewed by W. J. Bell (1991). Animal locomotion has been analysed in detail by J. Gray (1968) and his co-workers. Subsequent research has confirmed the basic principles already established. Taylor (1989) has reviewed the locomotor adaptations of Carnivora.

The evolutionary effect of selection by predators on their prey is outlined in Chapters 3 and 4. Less marked, but nevertheless present, is the effect of selection on predators for efficient predation – selection for speed or silence, aggressive crypsis and so on, which are considered in the present chapter.

Predatory animals of arid lands include arthropods, especially arachnids and insects, reptiles, birds and mammals. Of lesser importance are centipedes, which are only to be found on the fringe of the desert, and amphibians, of which there are but a few xerophilous species. Although many scorpions actively seek their prey (Cloudsley-Thompson 1981b), probably even more are sit-and-wait predators (McCormick and Polis 1990; Sect. 2.2), as are web-building spiders (Araneidae, Linyphiidae and Theridiidae) and crab spiders (Thomisidae). Most of the tarantulas (Mygalomorphae), including trap-door spiders, are also sit-and-wait predators. Among predatory insects and many araneomorph spiders, on the other hand, the majority of species hunt their prey actively. The families Lycosidae and Zodariidae are particularly well represented in arid environments, the former by wolf spiders of the genera *Evippa*, *Evippella* and *Evippomma*. These are speculatory predators that hunt their prey visually. The Zodariidae are often small unicolorous spiders active in sunshine with retreats beneath stones. Several species feed on ants. In desert regions there is always a conflict in diurnal forms between the need to forage in the open and, at the same time, to avoid excessive heating. This is usually achieved by shuttling between the desert surface and shelter in a burrow, or on a plant above the scorching sand.

Solifugae (Solpugida) have been observed in the field to exhibit rapid locomotion with frequent changes of direction. This type of random cursorial behaviour has been recorded in several species, both nocturnal (Cloudsley-Thompson 1977b) and day-active (Wharton 1987). When attacking scorpions, the adversary's tail is immediately grasped and the sting severed. This is merely an instinctive response to a nearby moving object, although it may give the appearance of intelligence (Cloudsley-Thompson 1991). Solifugae are extremely active, and most of their above-ground existence is spent in foraging for food or locating mates. Although Muma (1967) described much of their activity as "simply investigating their surroundings", Wharton (1987) interpreted such activity in *Metasolpuga picta* as the typical foraging behaviour of a cursorial predator. Unlike most scorpions, Solifugae provide excellent examples of arthropods that hunt for their prey and do not wait for it to come to them. The same can be said of spider-hunting wasps (Pompilidae), which scour the desert surface in an endless search for their spider prey. By means of their highly developed power of stinging, they are able to overcome even the largest of tarantulas (Theraphosidae) to stock the burrows in which their larvae develop. The family Sphecidae likewise contains a large number of solitary wasps that inject venom which paralyses the prey that their larvae later consume.

Amphibians respond to moving objects in one of two ways, turning towards them if they are small, and moving away when they are large. Detection of prey by adult anurans is largely visual, but aquatic larval forms also make use of the lateral line system and their sense of smell. In general, amphibians do not appear to be praticulary well adapted as predators (Wilczynski 1992). Most post-larval Anura have adopted a sit-and-wait strategy, while active foraging is more common among salamanders and caecilians. The latter are found only in moist tropics, whereas both Caudata and Anura occur in arid regions. All terrestrial amphibians, except caecilians, use their tongues for capturing prey. A sticky glandular secretion serves to make the prey adhere to the surface of the tongue. Post-larval salamanders, such as *Ambystoma tigrinum* (Ambystomidae) of the North American deserts, capture terrestrial prey by striking it with the posterior half of the tongue, while Anura enfold it with the lingual tip before withdrawing the tongue into the mouth (Duellman and Trueb 1986).

Most desert reptiles are carnivorous, feeding on a wide variety of prey. This may be either vertebrate or invertebrate, depending upon availability and size, and is usually found through active searching. Lizards deal with struggling prey by shaking it violently, or by striking it against the ground. Monitors (Varanidae) can probably break the backs of small mammals in this way. They also devour eggs when they can find them: these are swallowed entire and crushed by contraction of the muscles of the gullet. *Varanus niloticus* is known colloqually in Sudan as "the enemy of the crocodile" on account of its predilection for crocodile eggs. Some arid-zone lizards, such as *Agama agama*, are fairly omnivorous, feeding on flower petals, grass, dead leaves, fragments of groundnuts and other vegetable matter, as well as upon ants and termites (Cloudsley-Thompson 1981a). They become more carnivorous as they grow older, and adults have occasionally been observed to feed cannibalistically on their own young. *Uromastyx* spp. are similarly omnivorous, but become more markedly vegetarian when they grow older. In contrast, *Moloch horridus* and *Phrynosoma plathyrhinos* are both specialized predators of ants (Sect. 1.2). Slow-moving creatures such as these would be particularly vulnerable were they to rely upon flight to escape from their own enemies.

Whereas small lizards often feed exclusively upon insects, larger species may eat smaller lizards. Again, young snakes, as well as snakes of smaller species, prey on insects and arachnids, in contrast to larger snakes, which feed mainly or exclusively upon vertebrates. Small desert snakes are usually generalised predators. Shovel-nosed snakes (*Chionactis occipitalis*: Colubridae), for example, like most other denizens of the desert, are nocturnal, foraging in the open for insects and their pupae, spiders, centipedes and especially scorpions (Sect. 2.2). In contrast, the viper, *Bitis peringueyi*, lurks beneath the sand and rises up to engulf its victims – lizards such as *Aporosaura anchietae* (Lacertidae) – during the daytime. Unlike most other snakes, its eyes are situated on the top of the head so that it can be almost completely buried and yet retain a full field of vision. A sit-and-wait predator (Sect. 2.2), it remains with most of its head and body beneath the surface, where the sand remains relatively cool as it waits in ambush for lizards and other small vertebrates, including birds (Louw 1972; Louw and Seely 1982).

Desert reptiles show a variety of locomotory patterns which are used both for obtaining food and taking shelter. Adaptions for burrowing are often not particularly marked, apart from the possession of nasal valves, which prevent soil grains from being inhaled; but, in sand-swimming desert lizards and snakes, the nose or rostrum is frequently pointed or shovel-shaped. The nostrils may be directed upwards instead of forwards as a protection against the entry of sand. The eyes, nostrils and mouth can be closed by valves when the animals dive head first into loose sand. Their bodies are covered with smooth scales which cause little friction, and the legs of such lizards may be reduced, or even lost, so that locomotion is accomplished entirely by wriggling. Amphisbaenians, too, are legless burrowers with very smooth scales (Buxton 1923; Stebbins 1943; Bellairs 1969; Pough 1969; Huey et al. 1974).

Lizards have evolved fringes of elongated projecting scales on their toes at least 26 times and in seven different families. The morphology of these fringes varies according to the type of substrate. Species that run on wind blown sand usually have triangular, projectional and conical fringes, while riparian species that run on water tend to have fringes whose shape varies from narrow to wide rectangles (Luke 1986; Bauer and Russel 1991).

Snakes can move about in several different ways (Guibé 1970), of which serpentine locomotion is the most common. In this, the propulsive force is derived from the thrust made by the curves of the body against projections from the ground such as stones, plant stems and other irregularities. Another method of locomotion is concertina movement, again pressing the coils of the body against irregularities.

The sidewinding locomotion of vipers and rattlesnakes is especially efficient when moving over smooth, sandy surfaces. It enables short-bodied snakes, such as the Egyptian asp (*Cerastes cerastes*), the Namib desert viper (*B. peringueyi*) and the American desert rattlesnake (*Crotalus cerastes*) to move with considerable speed, yet keeping much of the body from making contact with the hot surface of the ground. Finally, pythons, boas and large vipers are able to creep forward by rectilinear locomotion with the body extended in an almost straight line. The broad, ventral scales are raised, drawn forward, placed on the ground, and the rest of the body then pulled after them (J. Gray 1968; Gans 1970).

In addition to its obvious advantage in allowing a snake to cross hot desert sand without overheating, sidewinding is quicker than other forms of locomotion on a loose substrate. Furthermore, it gives a misleading impression of the direction in which the snake is moving, so that it can surprise its prey with its unexpected speed. Sidewinding was first clearly described by Mosauer (1932) and subsequently analysed in detail by Gans (1970, 1974). A sidewinding snake achieves firm static contact by moving so that its body lies almost at right angles to the direction in which it is traveling, and its track in the sand appears as a series of parallel lines each at an angle of about 60 degrees to the snake's direction of movement (Fig. 1).

Although most desert snakes are flesh-eaters, a few small fossorial forms, as we have seen, are insectivorous. These include the worm snakes (Typhlopidae and Leptotyphlopidae), which lead secretive lives under rocks and stones or in burrows underground. Their eyes are rudimentary and their dentition reduced in association with their specialized diet: they feed almost entirely on termites and other small, soft insects. Along with other non-poisonous desert snakes, including the Boidae and most Colubridae, worm snakes have no special method for killing their prey. They either suffocate it by constriction, crush it with their jaws, or swallow it alive so that it dies either from lack of oxygen or from the action of the digestive juices. Constriction has been perfected by the

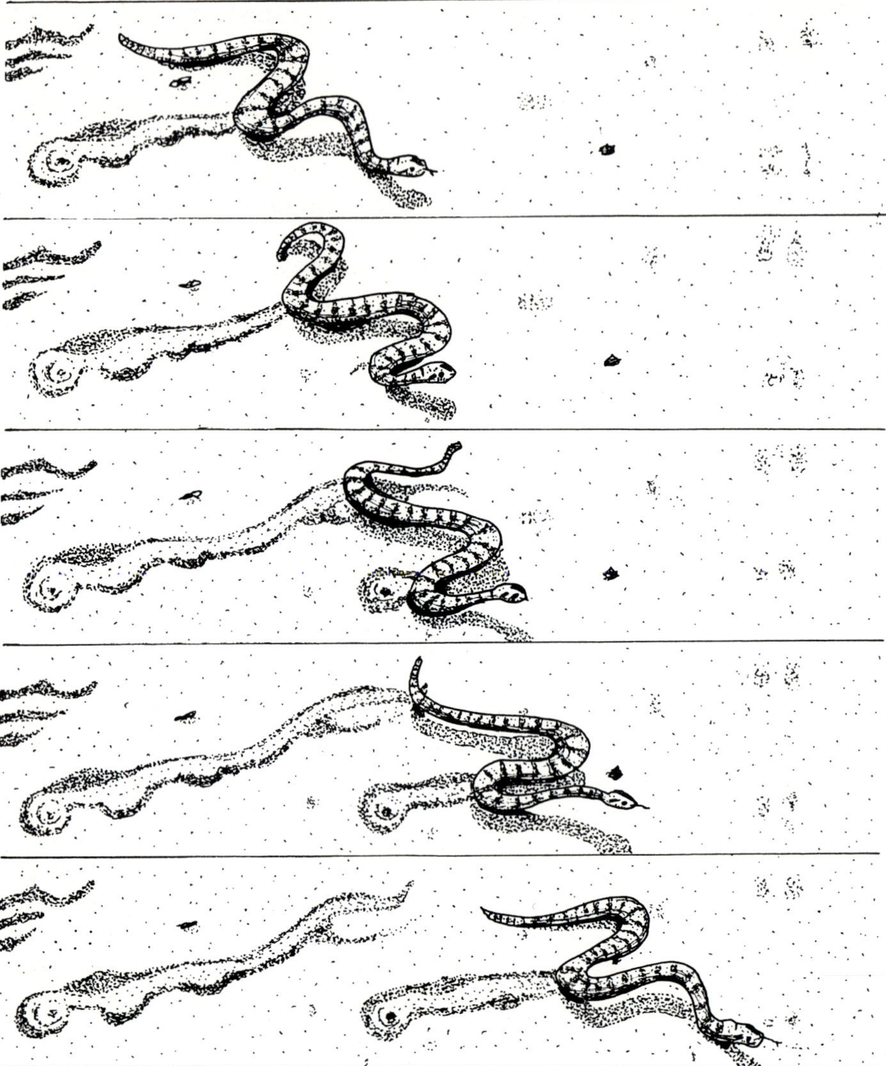

Fig. 1. Diagram to illustrate the principles of sidewinding. (After Gans 1974 in Cloudsley-Thompson 1994)

pythons and boas, but it is also employed by many Colubridae and snakes of other families that feed on warm-blooded animals. The prey is usually first seized by the jaws, then one or more coils are wrapped round it and contract, thereby stopping its heart from beating and preventing it from breathing. The process is also found in North American bull snakes (*Pituophis* spp.: Colubridae). King snakes (*Lampropeltis* spp.: Colubridae) use the same method for subdueing rattlesnakes, twisting their bodies many times round that of the victim (Bellairs 1969). Ontogenetic changes in prey preference are frequent.

Hinged teeth have evolved in at least three different lineages of snakes, apparently as an adaptation to feeding on hard-bodied prey, especially scincid lizards. These teeth fold backwards against the jaws rather than being firmly ankylosed: a hinge of connective tissue at the base of each tooth is associated with suites of cephalic modifications that enable the snakes to grasp and swallow hard-bodied prey. Folding takes place when forces are applied to their leading surfaces, but lock in an erect position when the forces come from behind – as would occur during the retraction movements of ingestion or when a prey item struggles to escape (Savitzky 1981).

Venom is but one of the many adaptations that contribute to the success of reptiles in predation. The only poisonous lizards are the Gila monster (*Heloderma suspectum*) and the Mexican beaded lizard (*H. horridus*) (Helodermatidae; Sect. 4.6.1). Ponderous and sluggish, they prowl at dusk, feeding on the eggs of other reptiles and of birds: they also eat lizards and small mammals when they can catch them (Cloudsley-Thompson 1994). Venomous desert snakes include rattlesnakes and other pit-vipers (subfamily Crotalinae) and Old World vipers (subfamily Viperinae). Those together comprise the family Viperidae, while the Elapidae includes the cobras, coral snakes, kraits and death adders (Bellairs 1969). The common death (or deaf) adder (*Acanthophis antarcticus*: Elapidae; Fig. 2) and its congeneric species are unique Australian elapids which have both the appearance and the behaviour of vipers. The heads are triangular in shape, while the tails are thin. *A. antarcticus* feeds upon lizards, birds and small mammals. Its poison is extremely toxic. It may be significant that the venoms of desert animals, such as scorpions and snakes, appear to be more powerful than are those of related species from more humid areas, but data on this matter are lacking. There is some evidence that prairie rattlesnakes (*Crotalus viridis*), and presumably other species, control the quantities of venom extruded in subduing their rodent prey (Hayes 1995).

Rattlesnakes and other pit-vipers of the subfamily Crotalinae possess a sensory pit on each side of the head between the eye and the nostril.

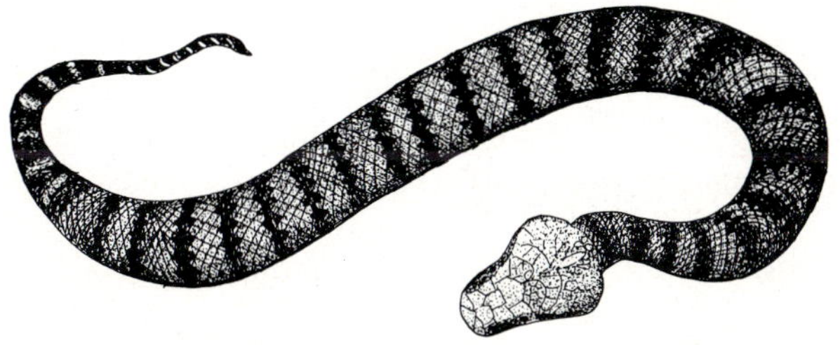

Fig. 2. Death adder (*Acanthophis antarcticus*). (Cloudsley-Thompson 1994)

This is a sense organ richly supplied with nerves. It responds to heat radiation and can detect thermal variations of less than 0.2 °C, enabling these snakes to find warm-blooded prey at night. After the prey has been struck and poisoned, it is followed by scent until it drops dead and can be swallowed. The sense organs concerned are known as the organs of Jacobson. They are present in chelonians, the tuatara, and are particularly elaborate in lizards and snakes. Each Jacobson's organ takes the form of a hollow dome, lying above the roof of the mouth and communicating with it through a narrow duct. Minute particles of scent are picked up by the tips of the forked tongue, and conveyed back to the organs of Jacobson. This explains the constant flicking movements of the tongue, seen in a prowling snake and monitor or other scleroglossan lizard (see Bellairs 1969; W. E. Cooper 1990, 1994a, b; Cloudsley-Thompson 1994).

Day-active snakes, especially those that are well camouflaged, tend to lie in wait (Sect. 2.2) until their prey comes close enough to be captured by a sudden lunge. Nocturnal species, on the other hand, tend to prowl around actively searching for prey in suitable places after nightfall.

The process of engulfing a meal depends upon the mobility of the bones of the upper jaw and the elasticity of the ligament connecting the two halves of the lower. While a firm grip is maintained with the teeth on one half of the jaw, those on the other side are relaxed and pushed forwards, the teeth being disengaged in the process. The upper and lower jaws are also worked alternatively, so that the prey is pulled forcibly into the mouth and forced down the throat while upper and lower jaws are dislocated (Bellairs 1969).

Specialized structures and behaviour are not required to grab and swallow an insect. The sandfish lizard (*Scincus scincus*: Scincidae) of the Sahara desert can respond and orient to unsuspecting insects moving over the surface of the sand at distances up to about 15 cm. The lizard obtains directional information from the vibrations engendered by its prey, localises the insects, and emerges from the sand to capture them. When walking on the surface, *S. scincus* often stops and plunges its head into the sand. This aids the detection of vibrations and, presumably, facilitates transmission of sand-borne vibrations to the inner ear (Hetherington 1989).

In some Sauria, the tongue has become adapted for rapid projection and capture of the prey. The rainbow lizard (*Agama agama*) and *Moloch horridus* , for instance, can protrude their tongues for a short distance to capture ants and other small insects which adhere to the sticky tips. The trait reaches its most extreme limit in chameleons (Chamaeleonidae) such as *Chamaeleo calyptratus* of the Negev, *C. chamaeleon* and *C. africanus* of the Sahara, *C. namaquensis* and *C. dilepsis* of the Namib desert, whose tongues can be projected to a distance equal to about one and

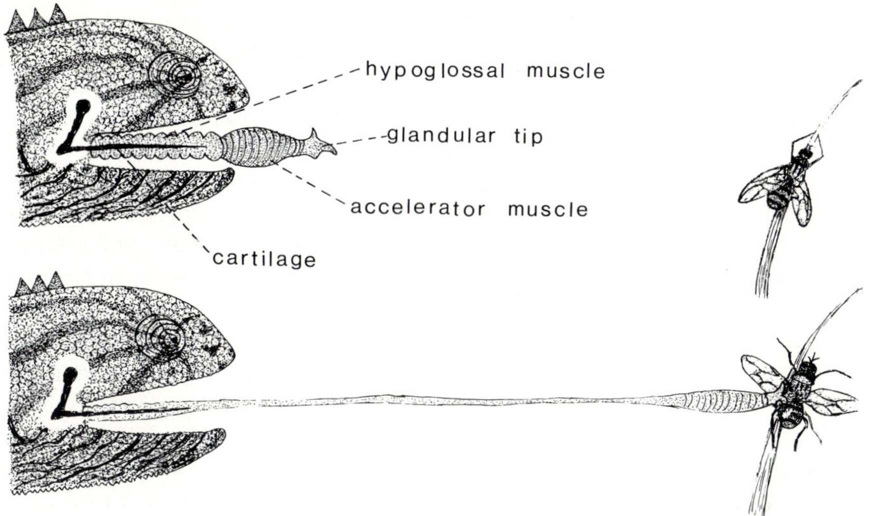

Fig. 3. Mechanism of extension of the chameleon's tongue. *Above* Tongue packed in tight pleats on a cartilage in the back of the mouth; *below* tongue projected forward by relaxation of the hypoglossal muscle and contraction of the accelerator muscle. The prey is trapped by the sticky glandular tip

a half times the overall body length, excluding the tail (Bellairs 1969). When retracted, the chameleon's tongue is in the form of a tube with walls consisting of longitudinal fibres of the hypoglossal muscle. These keep the tongue packed in tight pleats on a pointed bone or cartilage at the back of the mouth. The terminal, club-shaped portion of the tongue contains a powerful accelerator muscle which begins where the longitudinal fibres end. This runs radially in a transverse plane instead of being circular like the muscle of a sphincter. When contracted, the pressure engendered by the accelerator muscle is distributed hydraulically.

As they stealthily approach their prey, chameleons hold their heads forward, open their mouths, and move their tongues in front of the jaws. Contraction of the accelerator muscle moves the tongue off the central cartilage and, when the longitudinal hypoglossal muscle is relaxed, the tongue shoots forward like a released spring (Fig. 3). In addition to being glandular and sticky, the tip of the tongue has a small depression, the edges of which can be used to grasp the prey. Locusts and other insects that are too powerful to be trapped by adhesion alone are caught by larger chameleons in this way. The tongue is retracted more slowly than it is projected, especially when the prey is heavy.

Although individual species of reptiles undoubtedly have specific techniques for handling different types of prey, foraging behaviour can usually be divided into widely foraging and sit-and-wait (Sect. 2.2) predation. The first category may, in turn, be subdivided into cruising foraging and intensive foraging (Schoener 1971; Regal 1978; Huey and Pianka 1981; Heatwole and Taylor 1987). Life history and morphological characters correlated with foraging mode including camouflage, speed, streamlined body shape, autotomy, large brains and learning ability (Huey and Pianka 1981; Heatwole and Taylor 1987).

Sit-and-wait predatory lizards primarily use vision to detect larger, mobile prey, while active foragers and herbivores tend to use chemical cues for the detection of immobile, clumped prey (Rand 1994). In their important paper on the ecological consequences of the foraging mode mentioned above, Huey and Pianka (1981) postulated a correlation between the use of chemoreception for the location of prey, and wide foraging. Sit-and-wait predatory lizards in the Kalahari, Western Australia and North American deserts prey on more active animals, but eat less food than do widely foraging species. Foraging mode is related, moreover, to shape, physiological capacity and neurobehavioural phenomena, and its ecological correlates are extremly complex. According to W. E. Cooper (1994b), the strong association between mode of foraging and the presence or absence of the chemical discrimination of prey is consistent with the hypothesis that foraging ecology influences chemosensory behaviour. The extent to which this relationship might be the consequence of phylogenetic inertia rather than adaptation is not clear. Rand (1994), however, expressed surprise that when an active foraging lineage, such as the geckos, evolved sit-and-wait foraging, they should have lost the ability to locate their prey by chemical means.

Raptors attack a considerable range of prey species which require different degrees of agility for their capture. Thus, sparrowhawks, goshawks and falcons chase mainly other kinds of birds, whereas buzzards, eagles, vultures, kites and owls tend to prey on relatively slow-moving mammals for which they have to search. Species of the first group have small intestines which are 20–30% shorter than predicted on the basis of body size, and 50% shorter than those found among members of the second group of equivalent body size. It seems that some species have evolved alimentary tracts shorter than are necessary for maximum digestive efficiency and thereby enhance their efficiency in capturing prey (Barton and Houston 1994).

Not surprisingly, the whiteness of ostrich eggs is an adaptation which protects them against overheating – but at the cost of greater vulnerability to predators such as jackals, hyaenas and lions. Egyptian vultures (*Neophron perenopterus*) have developed the technique of throwing stones at ostrich eggs to shatter the shells – a trick that no other vulture species or other birds of prey have discovered. Each missile is hurled with a downward flick of the beak while the Egyptian vulture stands on the ground near the egg. It is not dropped from the air, and a number of direct hits are required to break the 2-mm-thick ostrich egg shell (Van Lawick-Goodall 1968). According to Bertram (1992), jackals probably break ostrich eggs by rolling them against each other, while lions and hyaenas use their teeth. However, H. Kruuk informs me (per. comm.) that he considers it impossible for jackals to break ostrich eggs, although they do roll them around. Spotted hayenas (*Crocuta crocuta*: Hyaenidae) likewise cannot open ostrich eggs; but brown hyaenas (*Hyaena brunnea*) have no problems with them (Mills 1990). In his book *Hyaena*, Kruuk (1975b) illustrates photographically an ostrich egg being rolled for about 8 m by a golden jackal (*Canis aureus*), then bitten by two spotted hyaenas, broken by an Egyptian vulture, which was almost immediately displaced by two hooded vultures (*Neophron monachus*), which would not themselves have been able to open it, and finally removed by another hyaena to eat in a quiet place.

Despite the fact that no less than 33 species of birds of prey, including vultures, are known to breed in deserts, not one is uniquely characteristic of that ecobiome. Some, such as the Palaearctic lanner (*Falco biarmicus*) and North American sooty falcon (*F. concolor*: Falconidae) take advantage of the peculiar circumstances of the desert to prey upon small migrant birds that are compelled to seek shelter in places where they can easily be found. Along with peregrines (*F. peregrinus*), they prey heavily on migrant songbirds, hunting these, doves and sandgrouse when they fly to drink at scarce waterholes (L. Brown 1976). Peregrine falcons are amongst the most highly evolved of raptors. They can even kill birds larger than themselves, such as bustards (Otididae), and almost all their prey is captured on the wing and in the air. Peregrines have exceedingly keen vision, and foraging individuals fly high in the sky over mainly open country, including the fringe of the Sahara. When potential prey is sighted below, the peregrine closes its wings and dives or "stoops", often from the direction of the sun so that it cannot readily be seen by its victim, and achieves a speed of nearly 300 km/h (nearly 100 m/s). Most prey are killed instantly by the falcon's talons. Larger prey are allowed to fall to the ground and are eaten there, while small items are carried away in the peregrine's claws (Pianka 1978).

Raptors adapted to the desert are of two main types – those derived from subtropical woodlands or steppes, and those characterisitc of drier tropical savannas but which also colonise deserts. Because so many prey animals are nocturnal, owls (Strigiformes) have a certain advantage over the day-active Falconiformes. Nevertheless, even naked rock desert south of the Atlas Mountains supports Bonelli's eagles (*Hieraaetus fasciatus*), peregrine falcons and Egyptian vultures. Despite the harsh climate, open desert country is favourable to predatory birds coming from less extreme environments, although they must have to vary their way of life to exploit new situations (L. Brown 1976). Indeed, desert raptors must be agile and use a variety of hunting techniques to obtain a wide range of food items. They sometimes cooperate in hunting prey, which they do not need to do in vegetated regions.

Owls have relatively large ear holes with flaps around them which help to channel sound waves into the ear, but they are hidden beneath the heat feathers. Owls are most sensitive to high-pitched sounds which are on the same wave lengths that their rodent prey use to transmit calls to their offspring and other members of the species. The calls they themselves make are of a lower pitch and carry for considerable distances. Their feathers are very soft and thick so that the sound of their own flight is muffled and the faint noises of the prey can be heard more closely. At the same time, the prey is not warned of their approach (Cott 1940). In many species, the ears are asymmetrical, of different sizes, and set at different levels on either side of the owl's head. Consequently, a noise coming from an angle will reach one ear a fraction of a second earlier than the other. It is this infinitesimal delay that indicates the direction of the source of the sound, just as does the delay in the arrival of ground vibrations to different legs of a waiting scorpion (Sect. 2.2).

Some birds make use of other animals to flush insects from the vegetation in which they would otherwise be hidden. In the arid parts of northern Africa, carmine bee-eaters (*Merops nubicus*: Meropidae) ride upon the backs of goats and other larger mammals (Cloudsley-Thompson 1965), white-bellied storks (*Ciconia abdimii*: Ciconiidae), or kori bustards (*Ardeotis kori*: Otididae), whence they swoop in pursuit of insects disturbed by their mounts (Curio 1976). Many small desert passerines, such as larks, wheatears, warblers, chats, and flycatchers, eat mainly insects, as do nightjars (Caprimulgidae) and swallows (Hirundinidae). All of them are adapted, in their own individual ways, for catching prey. Likewise, seed-eaters and other herbivorous birds are equally adapted for their particular diets. Nocturnal predators that hunt by sight may be constrained by the level of light intensity. This problem should be especially acute in aerial predators such as the poorwill (*Phalaenoptilus nuttallii*: Caprimulgidae). Comparison of its diet (assessed by analysing faecal samples) with insects taken in light-suction traps, suggests that poorwills are, indeed, constrained to take prey larger than a certain threshold. Above this, however, particular types and sizes of prey may be selected (Bayne and Brigham 1995). The food of desert birds is discussed by Maclean (1996), while a forthcoming book in this Series will be devoted to the adaptations of avian desert predators.

Carnivorous mammals of arid regions may be small, such as the insectivorous grasshopper mice (*Onychomys* spp.) of North America, or large, such as lions (*Panthera leo*), leopards (*P. pardus*) and hyaenas (*Hyaena brunnea* and *Crocuta crocuta*) of the Kalahari and Namib and *H. hyaena* of the Sahara and Arabian deserts. Many of them hunt by speculation (Curio 1976), searching in likely places for their prey, which is detected visually or by scent and sound. Desert carnivores have catholic diets and may evolve specific methods for dealing with awkward prey. For instance, some mongooses throw bird's eggs or millipedes against rocks to smash them (Eisner and Davis 1967). The African white-tailed mongoose (*Ichneumia albicauda*), however, holds eggs between its front paws, nibbles a hole in the top of the shell and daintily licks the contents with its tongue (J. L. Cloudsley-Thompson, unpubl. observ.)

According to Bertram (1975), group living and communal hunting may lead to co-operation (e.g. in lions; Sect. 7.1). This is advantageous to some predators because it improves their chances of locating food. At the same time, it may enable others, such as hyaenas and wild dogs (*Lycaon pictus*), to capture larger prey than they would otherwise be able to overpower, and results in individuals of a group competing better than would solitary individuals. On the other hand, arid regions in which food is in short supply cannot support large prides of lions or clans of hyaenas. An illuminating example of limitation to the size of a group is provided by a small troop of chacma baboons (*Papio ursinus*) living in the dry Kuiseb river canyon, Namib desert. These forage for animal prey and plant matter away from free water for periods of up to 5 days, but the harsh environment and social pressures contribute to an extremely poor infant survival rate (two infants in approximately 8 years) and, consequently, the size of the troop is much smaller than that of chacma baboon troops in less arid regions (Brain 1990).

Other predatory mammals of arid and semi-arid lands include genets (*Genetta genetta*), mongooses (*Herpestes* spp.), meerkats (*Suricata suricatta*), foxes such as the Old World fennec (*Fennecus zerda*), sand fox (*Vulpes pallida*), silver fox (*V. chama*), Rüppell's fox (*V. rüppelli*), and the North American kit fox (*V. macrotus*), jackals (*Canis aureus*), wolves (*C. lupus*), cayotes (*C. latrans*), bobcats (*Lynx rufus*), caracals (*Felis (Lynx) caracal*), sand cats (*F. margarita*) and so on. Most, if not all, of these predators hunt their prey by speculation and tend to take a variety of species. The fennec, in particular, is more catholic in its diet than other foxes and lives mainly on insects, dates and so on. Indeed, its liking for sweet things probably accounts for the otherwise inexplicable fable of the fox and the grapes. Many large carnivores, such as lions and hyaenas, are also scavengers and will be discussed below (Sect. 2.3).

Techniques for the capture of prey by mammals may be complex. For instance, the southern grasshopper mouse (*Onychomys torridus*) is a nocturnal predator of the North American deserts, where it is able to overcome beetles that spray noxious chemicals, venomous scorpions, and even to capture lizards, birds and mice larger than itself. For each prey species it adopts a different technique. Small, undefended arthropods are usually restrained by leaping on them and biting their heads. Tenebrionid beetles (*Eleodes* spp.) are grabbed, held upright and the quinone-spraying rear end (Sect. 4.6) thrust into the sand, where it is discharged safely, while they pursue *Mastigoproctus giganteus* (Thelyphonida; Sect. 4.6.2) until its spray has been exhausted (Eisner and Meinwald 1966). Scorpion tails are immobilised prior to further attack

(Horner et al. 1965), while mice and lizards are either strangled or else quickly killed by a bite through the cranium or spinal column (Egoscue 1960; Horner et al. 1965). When attacking the potentially dangerous grasshopper *Brachystola magna*, *O. torridus* alters its mode of attack and crushes the insect's spiky and powerful hind legs at the femoral-tibial joint before biting the head. In this way, the prey is immobilised and its potential weapons eliminated (Whitman et al. 1986). Grasshopper mice are effective, resourceful predators, incorporating a variety of techniques for tackling diverse prey. Such optimally adaptive predatory behaviour has been suggested by Langley (1981). Whether these techniques are learned or innate is not clear, but they certainly enable a wide variety of prey to be overcome and consumed.

In general, carnivores prey on herbivores or other carnivores that are smaller or approximately the same size as themselves, unless they cooperate as a pack or pride. It is not surprising that natural selection should operate in favour of individuals that remain within the limits of their capability. The attenuated food resources of desert-dwelling species cannot always support pack hunters such as wolves and wild dogs which are able to kill prey much larger than themselves. Although lions, leopards and other carnivores are demonstrably capable of killing animals very much heavier than themselves, this does not invalidate the claim that their normal prey is seldom more than twice their own weight (Kruuk and Turner 1967; see discussion in Harvey and Gittleman 1992).

Different families of large terrestrial carnivores have evolved typcial hunting techniques to which they are morphologically adapted – dogs (Canidae) to swift and prolonged running which demands great stamina; cats (Felidae) to a concealed approach followed by a quick dash. Hyaenas (Hyaenidae) are coursers like dogs, killing and scavenging on a wide variety of prey animals. They have extremely strong teeth and jaw muscles for crushing bones (see below). Contrary to earlier belief, it is now thought that carnivores tend to hunt in groups because they live in groups and not because they necessarily hunt cooperatively or experience greater foraging returns per capita than solitary individuals do (Caro and Fitzgibbon 1992). Nevertheless, the chances of escape are reduced when a large prey animal, such as a wildebeest, is surrounded by a group of wild dogs or hyaenas. Prey-catching by mammals has been reviewed by Ewer (1968, 1973), and foraging behaviour in animals surveyed by Kamil et al. (1987).

2.2 Ambushing and Disguise

There can be no clear distinction between actively searching for prey and ambushing it. Many desert predators move towards their prey, or to places where it is liable to appear, and there lie in wait for it. They exhibit a combination of hunting by speculation and lying in wait.

Ant lion larvae (Myrmeleontidae) provide a classic example of desert insects that ambush their prey while they lie buried at the bottom of their pits, leaving only the large jaws protuding. An ant or other insect wandering over the edge of the pit slides to its doom, assisted by particles of sand that the ant lion jerks towards it. Fly larvae of the rhagionid genera *Vermilio* and *Lampromyia*, commonly known as "worm lions" (Wheeler 1930), also construct conical pits for the capture of their prey. Somewhat similar tactics are employed by larvae of the Australian carabid beetles *Sphallomorpha colymbetoides* and *S. nitiduloides*, which prey on meat ants (*Iridomyrmex purpureus*: Dolichoderinae). They dig burrows near the nests and paths of the ants. When a worker approaches, the larva lunges and grasps the ant's legs. After its struggles cease, the carabid sucks out the haemolymph of the ant and discards the shrivelled corpse (Moore 1974). As in the case of ant lions, several *Sphallomorpha* burrows are often to be found grouped together, which may improve the predators' hunting success because captured ants discharge alarm pheromones which attract other ants from the colony. The calliphorid fly *Bengalia peuhi* (Fig. 4) and its rare congenor *B. minor* are visual predators of ants in the northern Sudan. These marauding insects snatch both larvae and adults of the small *Monomorium salomonis* (Myrmicinae) with their raptatorial sucking proboscis (Cloudsley-

Fig. 4. *Bengalia peuhi* preying on ants (*Monomorium salomonis*). (Cloudsley-Thompson 1963)

Thompson 1963). In fact, a diversity of insect predators settle near ant trains and prey on the ants, employing a variety of hunting techniques (Hölldobler and Wilson 1990).

Mantidae are rapacious predators in many arid regions. They sit on plants or trees, keeping a look out for insects which are trapped as soon as they fly within striking range of the fore legs. These are highly adapted: the coxae are elongated and mobile, while the femora bear thick spines and are grooved along their lower side. The tibiae, which are also spiny, fit into the groove. Just before their apex, which is usually produced into a hook, is inserted a reduced five-segmented tarsus. Praying mantids – so named because their raised fore legs with pinching segments partly open give the appearance of supplication – are variable in shape and highly camouflaged so that they resemble the vegetation on which they are poised. Those that simulate flowers have the advantage of attracting insect pollinators within their reach. However, unlike some auhorities, I believe the function of the protective resemblance (Sect. 3.5) to be mainly defensive rather than offensive. Species that look like twigs perfect their camouflage by holding their front legs in an extended position, but they must flex them before striking. By moving slowly with a swaying gait, however, as though shaken by the wind, they adopt a strategy likely to fool both predators and prey (J. Owen 1980).

Most desert scorpions are sit-and-wait or ambush predators. They emerge from their burrows at night and then remain motionless nearby (Polis 1979). The relationship between sit-and-wait foraging strategy and dispersal in *Scorpio maurus* has been investigated by Shachak and Brand (1983). Insects and other prey are detected, when they pass nearby, through the vibrations set up in the sand. Prey moving within 15 cm of the scorpion are sensed by tarsal sense organs, located and captured in a single motion. Brownell (1977) and Brownell and Farley (1979a, b) showed experimentally that in the case of *Paruroctonus mesaensis* (Vaejovidae), both the direction and distance of a source of vibrations is determined so accurately that a single turn and forward movement are sufficient to capture an insect the size of a cricket. The prey is located by compressional and surface waves in the sand. The former spread away from the source of the disturbance in all directions and have motion along the axis of travel: surface waves move across the surface of the soil and are vertically polarized. Velocities are comparatively low – 95–120 for compressional and 40–50 ms^{-1} for surface waves respectively. For frequencies between 1 and 5 kHz, the specific attenuation factor (Q) for sand is 18. Compound slit sensilla on the basitarsal leg segments of the sand-dwelling scorpions respond to surface waves generated up to a distance of 50 cm, while tarsal sensory hairs respond to high frequency compressional waves in the sand. Direction of stimulation is assessed by sensing which of the legs is stimulated first. Delays as small as 0.2 s elicit accurate responses, while information from compressional waves enables the scorpion to judge the distance of the source of stimulation. This depends either upon attenuation of the signal, or on appreciation of the delay between the stimulation of the tarsal hairs by compressional waves and the stimulation of the slit sensilla by the slower-moving surface waves (Brownell and Farley 1979b, c; see also Root 1990).

Although scorpions are extremely sensitive to light, vision is not important in the detection of prey. Air movements produced by moving animals are, however, detected by the trichobothria (Le Berre 1979; McCormick and Polis 1990). Chemoreceptive sensilla on the tips of the chelae, the tarsi of the legs and the pectines release specific feeding responses. Although they search actively for their prey (Sect. 2.1), the vision of Solifugae seems to be little better than that of scorpions (Cloudsley-Thompson 1961b) and the prey is detected mainly by touch. The limbs are covered with long sensory hairs and trichobothria, which facilitate this.

Spiders have evolved numerous methods for ensnaring, ambushing and trapping prey. Orb webs enable a large area to be monitored while the spider itself remains concealed and motionless. Trap-door spiders (Ctenizidae), particularly common in Australia, wait in their underground burrows behind camouflaged doors through which they spring on any insect that trips the signal threads of silk radiating around the entrance to the burrow. Some species stretch out their front legs into which the prey wanders. Substrate vibrations are probably detected by slit sensilla and lyriform organs (Foelix 1982).

Crab spiders (Heteropodidae and Thomisidae) lie in wait for their prey often in flowers or on vegetation, instead of chasing it as other hunting spiders do. Some species will spring at an insect with agility, but others remain in ambush until their stout front legs can grasp the prey firmly whilst the chelicerae are buried in its body (Henschel 1994). Certain species are able to change colour to match their background, but the function of this is probably more one of crypsis (Sect. 3.4) than of concealment for offence. Agelenidae construct funnel-shaped cobwebs consisting of a triangular sheet with its apex rolled into a tube in which the spider waits for its prey. These tubes are often found in the cracks of cotton soil (cracking clay) and in rock crevices. The threads of the web are not adhesive, but trip passing insects, which fall onto the sheet. Before they have time to recover, the owner of the web has darted from its tube and gathered them in.

The orb webs of Araneidae are usually attached to bushes. Those of *Argiope lobata*, a common inhabitant of sand dunes in the Mediterranean region, and of *A. sector* in Sudan are extremely large. *A. lobata* has an irregular shape and sandy colour which render it extremely inconspicuous to prey and predators alike (Lubin 1986). Scarcity of vegetation limits the numbers of Araneidae and other web-spinning families in arid regions. Surprisingly, the buckspoor spider *Seothyra henscheli* (Eresidae; Sect. 3.2.1) of the Namib constructs across the surface of the sand a web which traps prey without itself becoming covered with sand grains.

The web of *S. henscheli* consists of sticky silk lining the edges of a horizontal mat on the sand surface. The spider sits in a silken-lined burrow attached to the mat and, when arthropods become entangled in the sticky silk of the mat, they are attacked by the spider and pulled into its burrow. *Seothyra* burrows average 13 cm in depth during the hot season, at which depth the temperature fluctuates between 30–40 °C – well below that at which the spider experiences thermal discomfort (49 °C). By moving between the hot surface mat and the cooler burrow, spiders are able to forage at web temperatures above 65 °C. They are foraging at the thermal limit and capture their prey in significantly less time at surface temperatures above 49 °C than below (Lubin and Henschel 1990). Heatwole (1996) discusses in detail the sit-and-wait ambushing of prey by desert invertebrates.

Many desert amphibians and reptiles are cryptic sit-and-wait predators. Lizards, expecially, are usually cryptic for offence as well as defence (Sect. 3.4) and are not easily seen by their prey. Some wait until an unsuspecting insect settles near them and then capture the prey with a short swift dash, others use a combination of active foraging and lying in wait.

Fig. 5. Western diamondback rattlesnake (*Crotalus atrox*) showing conspicuous tail (Sonoran desert)

Like other sidewinding vipers, the cryptic *Bitis peringueyi* lies buried in the sand for hours on end, with only its eyes visible, waiting for an unwary prey item, such as a lizard, to pass within striking distance (Louw and Seely 1982; Lovegrove 1993). According to Curio (1976), the tail is moved if a small bird is spotted. When the latter approaches, motivated by curiosity, it may come within striking range of the snake's head. It seems quite possible that the conspicuous rattle of the rattlesnake (Fig. 5) may have evolved for a similar function and later developed into a warning signal (Sect. 3.8). Some geckoes, likewise, twitch their tails after prey has been detected and thereby attract it. Not only are sidewindling snakes able to attract prey while lying in wait but, because they appear to be moving in another direction, they are able to approach their prey unobtrusively (Sect. 2.1).

Birds cannot be said to ambush prey, but an attacking raptor will often take its prey by surprise. As already mentioned, falcons are said sometimes to stoop from the direction of the sun so that they are not noticed by their prey. Likewise, mammals that search actively for prey often approach it stealthily and attack suddenly from close range. A lion may trot towards its victim and speed up only when the latter has shown itself to be inattentive. Various ambush predators position themselves so that they can see their prey silhouetted against a light background. For instance, nightjars (*Caprimulgus europaeus*: Caprimulgidae) dart upwards to snatch insect prey that contrasts against the pale sky. Ambush predators are often markedly versatile in selecting optimal sites (Curio 1976).

Even large carnivores may increase their hunting success as a result of vegetative cover (Bothma et al. 1994) and their cryptic coloration. According to Lovegrove (1993), where dwarf shrubs (*Rhigozum trichotomum*: Bignoniaceae) dominate the vegetation on the banks of the Nossob River in the Kalahari, the black and straw-coloured manes of the lions blend superbly with the shrubs and golden grasses. When stalking, with belly and head close to the ground, lions and lionesses are extremely difficult to distinguish and, not surprisingly, solitary males can ambush even the most cautious adult gemsbok (*Oryx gazella*) although they usually let the females do the hunting. Not only are the leopard's spots an obvious aid to concealment in trees, their preferred daytime resting place, but they have a cryptic effect even in exposed desert shrubland. Throughout the deserts of Africa and Asia in which they occur, the ground colour of the fur of leopards matches the prevailing colour and tones of the habitat.

2.2.1 Aggressive Mimicry

Aggressive mimicry is sometimes employed by stalk and ambush predators, enabling them to deceive their prey until the final assault. It might be assumed that aggressive, or peckham's mimicry (first described by E. G. Peckham in 1889), as it is sometimes called (Curio 1976), is completely different from concealment or camouflage but, according to Wickler (1968), both concealment and mimicry of all types, except strictly mullerian (Sect. 3.6), are similar in that they involve deceiving the recipient of the signal. Nevertheless, they do involve quite different demands upon the animal that emits the signal. A number of examples can be seen among the faunas of deserts. For example, assassin flies of the genus *Bana* (Asilidae) in southern Namibia mimic bees and other Hymenoptera, which they pursue actively (Londt 1991). Possibly the flies benefit not only in being able to approach their prey without being recognised as predators, but the same time, no doubt, by batesian mimicry from potential predators that avoid venomous aculeate Hymenoptera. Pirate spiders (Mimetidae) use aggressive mimicry to capture web-building spiders. They produce vibratory stimuli on the webs of the latter, which induce the owner to attack what they mistake for prey. They themselves are then killed and eaten by the pirate spiders.

Ants form large colonies and are often abundant in the desert. Most species either bite, or sting, or both, and few vertrebrates eat ants regularly. It is not surprising, therefore, that ants provide models for many taxa of batesian mimics (Sect. 3.6) including mantid larvae, grasshoppers, Heteroptera and other insects. Several species of spiders from different families have evolved a close resemblance to ants (Cloudsley-Thompson 1995). Ant-mimicking jumping spiders (Salticidae) not infrequently feed upon their models. Other examples of predation on their hosts by ant mimics are afforded by *Seothyra henscheli* , which mimics the ant *Camponotus detritus* in the western and central Namib, whilst in the eastern Namib, where *C. detritus* is absent, males of *S. henscheli* mimic *C. fulvopilosus*, according to Curtis (1988; see discussion in Edmunds 1974).

Like cats, amphisbaenians and certain geckoes twitch their tails when they detect suitable prey. These movements distract the attention of the prey from the deadly head that is stalking it. Mention has already been made (Sect. 2.2) of this and the use by *Bitis peringueyi* of the tip of its tail to lure small birds within striking distance of the head. Several species of pit-vipers of the genus *Agkistrodon* from the moist tropics have conspicuously yellow or reddish tails whose brightness contrasts with the dull, inconspicuous colour of the body. When they are lying coiled at rest, the tail is hidden beneath the body but, if a lizard or frog approaches, the tail is raised vertically and wriggled, simulating a worm or caterpillar (Cloudsley-Thompson 1994). Aggressive mimicry has also been described among raptors such as merlins (*Falco columbarius*), which frequently fly in the fashion of small passerines when about to attack them (Curio 1976), and the zone-tailed hawk (*Buteo albonatus:* Accipitridae), which resembles the turkey vulture (*Cathartes aura*: Cathartidae; Willis 1963; Sect. 3.6).

2.3 Scavenging

Arthropod detritivores are dominant members of surface-active assemblages in most desert habitats, and are a major element of desert faunas generally. Most species forage on relatively mobile detritus at various times and seasons, and over landscapes with patchy and often ephemeral habitats. Convergence of assemblage structure occurs in similar habitats in different deserts or different parts of the same desert, and both competition and predation play a part in structuring macrodetritivore assemblages (C. S. Crawford 1991a). Animals that feed mainly on dead animals, or mainly on the dung or the exuviae of larger animals, are termed scavengers. Heatwole (1996) discusses scavenging by desert ants in some detail.

Scavengers form an important link in desert food chains. Many tenebrionid beetles are omnivorous and will scavenge on dead arthropods of various taxa or dung: adult *Blaps* spp. seem to confine themselves to a carnivorous diet. Competition with Diptera a well as with vertebrate scavengers must have had an important influence on the evolution of carrion beetles. The pre-adaptations of Diptera to scavenging include their extreme mobility as adults, and the rapid development of their larvae. In contrast, beetles are considerably less mobile, but live for very much longer – more than 5 years without feeding in the case of adult *Blaps requieni* (Tenebrionidae; Cloudsley-Thompson 1964). *Pimelia grandis* likewise feeds readily on dead insects and scorpions, but can withstand long periods of starvation. According to Hafez and Makky (1959), *Adesmia bicarinata* (Tenebrionidae) makes use of all possible sources of food encountered in the desert, including dry and fresh plant material, living larvae, and possibly human excrement. Pierre (1958) analysed the food of desert insects in some detail.

The main advantages to Diptera of their higher mobility is that they usually reach a cadaver first, and there is a greater chance for their larvae to complete development before the limited food resource is consumed. Futhermore, larval Calliphorinae metabolise large amounts of free ammonia, which is toxic to carrion beetles. In contrast, beetle larvae have the advantages of better-developed sensory and locomotory organs than those of fly larvae, and are better able to move a corpse and select the most favourable areas for feeding on it. Moreover, the adult beetles are often adapted for burrowing so that portions of carrion can be removed from areas accessible to flies to underground cells where the larvae feed (Crowson 1981). In general, carrion beetles show little specialization in respect to the type of carcass used, but tend to be restricted more by the types of habitat in which they operate. Moreover, many of the beetles regularly found on carrion – notably in the families Histeridae and Staphylinidae – are largely predatory, at least in the larval stage, and the favourite prey of most of them are dipterous larvae. By combating the pullulation of calliphorine larvae, such forms may considerably assist silphid carrion beetles (Crowson 1981). Many other carrion beetles, especially Dermestidae, evade serious competition with fly larvae by concentrating on carcasses in the later, drier stages of decomposition. Some feed on fur, tendons, hair and feathers. These represent extreme developments of the trend.

In a study of carrion in Australia (Borremissza 1957) the following stages of decay were noted: (1) First stage – mainly bacteria, Protozoa and roundworms; no beetles. (2) Putrifaction stage – fly maggots, mites and silphid beetles eating the decaying flesh. The predators of the maggots are Staphylinidae, Silphidae and Histeridae. Little ptiliid beetles were also found, probably feeding on the mites. (3) Fermentation stage, when the carcass is drying out. Skin and ligament feeders predominate – Dermestidae and Cleridae. (4) In the last, very dry stage, the only carrion feeders left are moth caterpillars, mites and a few small predatory beetles (Staphylinidae and Carabidae).

Some coprophagous beetles (Scarabaeidae) mould relatively enormous balls of dung which they roll away for their own benefit and that of their future larvae. The best-known exponent of the art is the sacred scarab beetle (*Scarabaeus sacer*) of Egypt. The female of this species detaches a portion of dung and forms it into a pellet, sometimes as large as a fist; she compacts this by pushing it backwards uphill with her hind legs and allowing it to roll down again. The ball is then rolled away by the beetle, who pushes when necessary with her broad head, or walks backwards, dragging it with her front legs. Not infrequently she is assisted by a friend, who is usually of the same sex, until a suitable place is reached and the owner commences to excavate a chamber for the reception of her prize. Sometimes the friend takes this opportunity to roll the ball away for her own use. If no such disappointment occurs, however, the *Scarabaeus* buries it in the subterranean chamber and remains with it until the food is entirely exhausted, when a fresh supply is sought.

In the autumn, a larger subterranean chamber is formed to which the beetle carries dung until it has accumulated a mass of provender the size of an orange. In this, an egg is deposited. In certain other dung beetles, such as *Copris hispanus*, the male and female combine to excavate an even larger earthen chamber containing from two to seven ellipsoidal balls of dung, in each of which an egg is laid. These dung balls are guarded while the larvae are devouring the food thus provided and, when the young beetles emerge, they are escorted to the exterior and the little family disperses (Crowson 1981). The devotion of the parent beetles to their young (Sect. 7.1) may well be the origin of the charming Arab proverb: in the eyes of a mother dung beetle her young are as the gazelle.

The larger scavengers of arid regions include kites and vultures, jackals, hyaenas and lions. According to L. Brown (1976), the following species of Accipitridae are resident in deserts: letter-winged kite (*Elanus scriptus*), Egyptian vulture (*Neophron percnopterus*), European griffon (*Gyps fulvus*), Rüppel's griffon (*G. ruppellii*), Cape vulture (*G. coprotheres*), hooded vulture (*Neophron monachus*), lappet-faced vulture (*Aegypius tracheliotus*) and white-headed vulture (*A. occipitalis*); but other species may sometimes penetrate far into the desert. *E. scriptus* is Australian, the remainder African or Eurasian.

Vultures compete both among themselves and other animals for shares of the same dead animal (Fig. 6) (Sect. 7.1.5). No detailed investigation of New World vultures has been made, but Kruuk (1967) carried out an important study of six Old World species occurring in East Africa. He found that they form three species pairs, each with different functions at the carcass. The small Egyptian and hooded vultures have thin bills and light skulls, suited to picking up scraps or stripping shreds of flesh from narrow species which the heavier beaks of larger species cannot reach. The African white-backed vulture (*Gyps africanus*) and Rüppel's griffon are large, with heavy bills and long necks, but relatively

Fig. 6. Griffon vultures (*Gyps fulvus*) dispossessed by a pi-dog (Thar desert)

light skulls. They are adapted for thrusting their heads deep into the bodies of dead mammals and tearing off soft flesh. Finally the lappet-faced and white-headed vulture are even larger, with heavy skulls and very heavy beaks, adapted for tearing tough flesh from bones and eating skin. Lappet-faced vultures can also open a carcass when no other species can. As a result of convergent evolution, New and Old World vultures look remarkably similar although they are not closely related.

Marabou storks (*Leptoptilos crumeniferus*: Ciconiidae) are important scavengers of the semi-arid areas of Africa, occasionally penetrating into true desert. They frequent carrion in company with vultures, feeding on virtually any animal matter from termites to dead elephants. On rubbish dumps they not only clean up scraps, but kill rats that would otherwise infest them (L. H. Brown 1982). The adjutant-bird (*L. dubius*) of India is closely similar in appearance and behaviour. These birds illustrate the point that there is no clear distinction between predation and scavenging. Nor, for that matter, is there a marked distinction between scavenging and kleptoparasitism (Sect. 5.3); or between insect predators and parasitoids (Sect. 5.4).

Spotted hyaenas (*C. crocuta*) were at one time thought only to be scavengers, unlike the related aardwolf (*Proteles cristatus*: Hyaenidae), which feeds on termites and, in the Serengeti, almost exclusively upon *Trinervitermes bettonianus*, according to Kruuk and Sands (1972). *C. crocuta* is often attracted to carcasses by vultures which are day-active – hence it is more often seen scavenging than hunting. At the beginning of the rainy season, large flocks of white-bellied storks (*Ciconia abdimii*) alight on the plains of Serengeti on their southward migration from Europe and North Africa. The hyaenas come running up, fooled by the storks; but they only do so once – they soon learn to discriminate between storks and vultures! (Kruuk 1975b). Eloff (1964) was the first to describe a situation in which spotted hyaenas were predominantly hunting predators that chased their quarry in packs. He deduced this information from tracks in the sand of the Kalahari desert. Here, packs mostly chase gemsbok, but also hunt blue wildebeest (*Connochaetes taurinus*), eland (*Taurotragus oryx*) and springbok (*Antidorcas marsupialis*). Pack hunting takes place almost exclusively at night when larger prey are selected than during the day (Kruuk 1966, 1972, 1975a). In northern Botswana, the prey of spotted hyaenas consists primarily of antelope, warthogs and zebra foals, and three quarters of their food is acquired by hunting. Clans of up to 50 hyaenas hunt in groups of two to four. Once a kill has been made, other members of the clan converge to feed. Hunting is carried out mainly at night, but also takes place on cool days. The hyaenas dash up to their prey in fan formation, usually downwind. Although capture rates are low, hyaenas are extremely persistent (S. M. Cooper 1990). In the Kruger National Park, they frequently scavenge the carcasses of large ungulates, especially buffalo, but more than half of their food comes from their own kills of mainly mediun sized ungulates such as impala, kudu and warthog, which become vulnerable under conditions of drought (Henschel and Skinner 1990). Again, spotted hyaenas kill from more than 50% up to as much as 98% of their food in Etosha. Springbok are hunted most frequently and, occasionally, zebras; but, unlike in other areas, hyaenas are minor predators of young zebras there (Gasaway et al. 1991).

Two species of hyaenas live alongside one another in Kalahari. One, the brown hyaena (*Hyaena brunnea*), is a solitary opportunist, the other, the spotted hyaena a gregarious specialist. *H. brunnea* is much better adapted to desert conditions, with its large feet and long, shaggy coat, than is *Crocuta crocuta*. The brown hyaena is also smaller than the spotted species and the sexes do not differ much in size, whereas spotted hyaena females are larger than the males. Much of the brown hyaena's food consists of small pieces of bone, odd legs, skulls and so on which are consumed alone. Almost exclusively a scavenger (94% of biomass) it also eats tsama melons (*Citrullus lanatus*: Cucurbitaceae) and gemsbok cucumbers (*Acanthosicyos naudianus*: Cucurbitaceae). Although these have low calorific values, they have a high moisture content and are rich in trace elements and vitamin C. They provide an important source of moisture for brown hyaenas, much of whose vertebrate food is dry when eaten. In contrast, the spotted hyaena feeds mainly on the calves of large antelope, especially gemsbok, most of which are killed (70%) rather than scavenged (Mills 1990).

When compared with the food of spotted hyaenas elsewhere, the fascinating flexibility of the ostensibly specialist species becomes apparent. Unfortunately, it is not possible to make a detailed comparison between the behaviour of the species in the Kalahari desert, Ngorongoro Crater and Kruger National Park, because different workers have not used the same methods to measure, for instance, home ranges and territories. Nor, indeed, has this always been possible. The prey of hyaena clans is not infrequently scavenged by lions – themselves at one time regarded as being strictly predatory. Most predatory carnivores are themselves scavengers except, perhaps, for cheetahs (*Acinonyx jubatus*). The balance between competing scavengers and predators is delicate, and a carcass may sometimes change ownership several times (Kruuk 1972; Schaller 1972; Curio 1976; Mills 1990).

More than any other large carnivores, however, hyaenas are adapted to scavenging. Their morphological and physiological features include bone-crushing teeth – the third premolars in the upper and lower jaws. The carnassial shear of the long blades of the upper fourth premolar and lower first molars is a modification for cutting through hide and tendon. The different teeth are used for specific purposes. Finally, hyaenas can digest even the largest pieces of bone and have the stamina to travel long at dusk and catch chironomids, mosquitoes and other aquatic insects. Just occasionally, an early bat flies out before the last of the kites has returned to roost, and falls into its clutches (Cloudsley-Thompson 1970).

Another good example is provided by Kruuk (1972), who pointed out that spotted hyaenas switch from scavenging from lions' kills in the day to hunting on their own at night. This is probably related to the fact that, during the day, hyaenas, like lions, observe kites and vultures descending while, at night, antelope and other potential prey are more vulnerable. In cases such as this, when predators switch their mode of feeding or prey at different times of the day, it is possible to assume that there is some causal connection between the two, resulting in the synchronization of activities. It is difficult, however, especially in the case of desert animals, to be certain whether a predator is active at any particular time, because its prey is only available or abundant then, or because both are influenced by some third factor or factors, such as climate. For instance, an annual drop in the numbers of smaller animals during the dry season in arid regions is probably related partly to seasonal food shortage, both animal and vegetable, and partly to the influence of unfavourable climatic conditions. The majority of insect species show life cycles adapted to exploit the short rainy season; the remainder of the year is passed in considerably reduced activity (Cloudsley-Thompson 1968). Similar seasonal cycles occur among scorpions and Solifugae in Sudan (Cloudsley-Thompson 1977b).

The times of activity of all plants and animals are determined by circadian "clocks". These are found in every cell of the body (Applin et al. 1987). and are synchronized by "master clocks" such as the hypothalamus of the vertebrate brain. In most desert organisms, they prescribe a nocturnal rhythm of activity, but some animals, including a few species of tenebrionid beetles, are day-active. Although climatic factors are important in determining time of activity, others, such as predation, foraging patterns and competition, also play a part. In different species and at different seasons of the year, any one of these may predominate. Both circaluna and circannual clocks are also involved. Perhaps more in deserts than elsewhere, the time and season of activity are crucial for survival.

3 Primary Anti-Predator Devices

Predation is one, if not the strongest, of all selective factors in most ecobiomes of the world. Because arid regions and deserts provide relatively little cover, the impact of predation is especially influential in these environments and, consequently, they provide some of the most extreme and dramatic examples of phenomena that are found almost everywhere on earth. Anti-predator devices can be subdivided into two main types, primary and secondary. Primary devices are defined as those which operate regardless of whether a predator is in the vicinity or not: they reduce the chance of an encounter between predator and prey, and include living in a burrow or hole, some forms of protective coloration, and the avoidance of detection by sight or sound (Cott 1940; Kruuk 1972; Edmunds 1974; Cloudsley-Thompson 1980; G. Costa 1995). Secondary anti-predator devices are called into play only when a potential predator is present. They will be analysed in Chapter 4.

3.1 Anachoresis

Many otherwise defenceless animals spend almost their entire lives hidden from predators in crevices, beneath bark or in holes in the ground. Such recluses are known as anachoretes (from the Greek word meaning "one who has withdrawn himself from the world"; Edmunds 1974). It must be remembered, however, that even anachoretes may need to emerge in order to feed or mate. Most of them are sit-and-wait predators and, like many spiders of the families Theraphosidae and Lycosidae, dart from their lairs to capture prey. In summer, male *Aphonopelma* sp. (probably *A. chalcodes*; Grammostolinae) leave their burrows and are to be found at night, wandering across the Sonoran desert in search of sedentary mates. Not infrequently also, they can be seen in daylight, especially after storms. At all times, however, crypsis may be valuable since enemies are never absent and light from the stars and moon in the clear desert sky is so bright that adaptive coloration is always important. A large proportion of desert animals are predatory, as we have seen (Chap. 1). At the same time, most predators are themselves the prey of other predators – for predation is especially severe in open country with little or no vegetation. They need to be able to recognise both their enemies and their prey.

Although comparatively large size in arthropods may be beneficial in reducing surface to volume ratios, at the same time it renders larger animals especially vulnerable to vertebrate predators. Indeed, the three factors of the desert environment that most influence its inhabitants are heat, drought and exposure to enemies. During their daily sojourn in retreats and burrows, nocturnal animals avoid the extremes of all these parameters. Nevertheless, they do not escape from them entirely, for the desert may still be hot at night, or it can be cold. It can be very dry, or even flooded, and predatory enemies are present at all times. Even anachoretes, buried deeply in dune sand, are liable to be dug up and eaten. This fate, for instance, befalls the insects and arachnids devoured by the sand-swimming Namib golden mole *Eremitalpa granti* (Insectivora), and the centipedes, scorpions, and beetle larvae captured by meerkats (*Suricata suricatta*; Fig. 12). Again the burrows of tarantulas (Theraphosidae) provide no security from tarantula hawks (*Pepsis* spp.: Pompilidae), which do not hesitate to invade them and secure their prey, while baboons (*Papio* spp.) regularly overturn quite large rocks as they forage for scorpions and insects.

3.2 Burrows and Retreats

Most small animals of the desert and arid zones behave as anachoretes for much of the time, and secrete themselves in burrows and retreats during the heat of the day, or during the dry season. These refuges also help to shelter them from the attentions of predatory enemies (Cloudsley-Thompson 1989, 1991). Solitary Hymenoptera comprise mainly burrowing forms. (The terms mining and boring are normally reserved for species that excavate galleries in the pith of plants). Some solitary wasps dig individual nests in the ground in which a single nest is placed; others dig communal nests or burrows with lateral chambers. Carpenter bees (*Xylocopa* spp.: Apidae) excavate long galleries in timber, or make nests in the stems of grasses and other plants. Despite this, they do not escape parasitism from Mutillidae (Sect. 5.4). Adaptations of desert arthropods and reptiles for burrowing have been discussed in an earlier book of this Series (Cloudsley-Thompson 1991) and will therefore not be considered here. It may, however, be worth mentioning that the giant cricket *Brachytrupes membranaceus* of the Namib desert unblocks the entrance to its burrow at dusk, comes to the entrance and stridulates. When disturbed, it breaks off its song and retreats into its hole. If no female appears within an hour or two, the male moves into the tunnel and closes the entrance with a plug of sand, using its head and fore legs.

This plug is removed the following sunset, and stridulation again takes place. If a female is attracted, mating takes place inside the burrow. The behavioral complexity of the species, which provides maximum security, is derived from its adaptations to life in an arid climate (G. Costa et al. 1987).

Many so-called burrowing animals do not burrow so much as push themselves into soil crevices. Geophilomorph centipedes, such as the large, subterranean *Orya barbarica* and *O. almohadensis* of North Africa, force themselves into the soil by elongation and contraction of the body. Telescoping is facilitated by the presence of intercalary tergites and sternites, which slide under the tergites and sternites in consequence of strong longitudinal musculature. The tendency towards a shortening of the segments and their increase in number reaches a peak in *O. barbarica*. In this species, the segments are at least 6 mm wide and only 1.5 mm long in the longitudinally contracted condition, and there are 107–125 pairs of legs. In species of *Geophilus* found under stones there are usually only 40–50 leg-bearing segments (Lewis 1981), while the speedy Lithobiomorpha, such as *Eupolybothrus cloudsleythompsoni* found alongside *O. barbarica* in the Tunisian semi-desert, and the Scolopendromorpha, are not adapted for burrowing.

Millipedes (Diplopoda) usually have more legs than most Chilopoda, and these likewise enable them to burrow by pushing themselves beneath rocks or into the soil. The day-active desert species *Orthoporus ornatus* and *Archispirostreptus tumuliporus* (Spirostreptidae), from the Chihuahuan and Negev deserts respectively, provide good examples (C. S. Crawford et al. 1987).

Spadefoot toads (*Scaphiopus* spp.: Pelobatidae) spend most of the year inactive in deep burrows, from which they emerge when rain falls (Sect. 5.3). They are quite unable to withstand the drought at other times of year and, even if they could, they would probably soon be eaten, despite their noxious skin secretions (Sect. 4.6.2). Most desert amphibians burrow (with the aid of well-developed tubercles on the hind feet) to a depth of about 30 cm, but some go down more than 80 cm (see Mayhew 1968).

Many desert reptiles spend most of their time in the security of holes and burrows when not actively hunting for food or on the lookout for mates. When both avoidance of predation and of excess temperatures operate synergistically, it is often impossible to distinguish quantitatively between the advantages afforded to each by burrowing. In the case of the sand lizard *Uma notata* (Iguanidae), however, the burrows are not deep enough to enable the animals to avoid high temperatures in midsummer; consequently, burrowing must be mainly a defensive device (Pough 1970). Burrowing by desert reptiles has been reviewed by Cloudsley-Thompson (1991), while, more recently, Arnold (1995) has analysed the effects of evolutionary history on the adaptions for sand-diving to avoid predators found among lizards. Once related lineages have diverged, differences continue to accumulate, even in simlar situations. Sand-diving also provides striking examples of convergence, and provides an illustration of how function can be used to polarize adaptive characters.

Most birds, apart from burrowing owls [*Speotyto (= Athene) cunicularia*] and Old World species of *Athene* , do not burrow; but African sand martins (*Riparia paludicola*) build their nests in long burrows, up to 1 m or more in length, which they excavate in steep banks or sometimes even in level ground. House martins (*Delichon urbica*) construct nests of mud collected in tiny pellets from damp ground. These take the form of half cups and are placed on cliffs or building. The African palm swift (*Cypsiurus parvus*) uses saliva to build its cup-shaped nest, which is stuck to the underside of palm leaves. These birds are a common sight flying over the Nile, especially at dusk. Birds nesting in holes in the ground include parrots (Psittaciformes), owls (Strigiformes), rollers (Coraciidae) and hoopoes (Upupidae), may of which inhabit semi-arid or arid regions, while bee-eaters (Meropidae) make their own burrows, often of considerable length, in banks or on flat ground. Several small passerines also nest in holes in the ground or in trees, while the tiny elf owl (*Micrathene whitneyi*; Fig. 7) of the hot Nearctic deserts, nests in holes in oaks (*Quercus* spp.) and cottonwood trees (*Populus fremontii*) in canyons. Saguaro cacti, however, provide their favourite nesting sites.

Fig. 7. Elf owl (*Micrathene whitneyi*) occupying the hole of a Gila woodpecker in a saguaro cactus

Fig. 8. Saguaro cactus (*Carnegiea gigantea*) showing nest holes of Gila woodpecker (Sonoran desert)

Like several other species of birds, they occupy the abandoned nest holes (Fig. 8) of the Gila woodpecker (*Centurus uropygialis*: Picidae). The poorwill (*Phalaenoptilus nutallii*) hibernates in rock niches during winter, and may return to the same hole year after year.

Almost all small desert mammals, apart from bats, are burrowers – even though some, including ground squirrels and meerkats (Fig. 12), may be day-active. Burrowing rodents such as gerbils, jerboas and kangaroo rats especially, are extremely successful desert animals, their holes being quite numerous in some areas. For most animals, a suitable substrate is necessary for the construction of permanent burrows. In regions of soft sand, they can either "swim" (Sect. 3.2.2), walk over the surface, or both. The golden mole *Erimitalpa granti* (Insectivora), for instance, walks for considerable distances on the sand surface in search of its anachoretic insect prey (Sect. 3.1) as do mole rat species such as *Batherygus suillus*, *Cryptomys hottentotus* and *Georychus capensis* (Rodentia) in search of food (Lovegrove 1993). The naked mole rat (*Heterocephalus glaber*) is almost completely anachoretic (Sherman et al. 1991; Sect. 7.1.1).

Many research workers have found the risk of predation to be a major influence on animal behaviour. In small desert rodents, predation constitutes an important foraging cost (J. H. Brown 1986), and may determine habitat partitioning among co-existing prey species. Hughes and Ward (1993) tested the hypothesis that the foraging activity of hairy-footed gerbils (*Gerbillurus tytonis*: Cricetidae) of the Namib desert decreases as risk of predation and costs of travel increase at greater distances from cover. They concluded that gerbils preferred to forage close to their burrows, regardless of the intensity of the moonlight. They were more active on nights of new moon, and decreased their foraging distance when the moon was full. Preference for foraging close to cover occurs at all times, indicating that the risk of predation is always high in open areas. Once gerbils make the decision to move out from cover (0–6 m) or away from their burrows, however, foraging costs do not increase significantly with increasing distance. Depending on the distance travelled, it is often advantageous to minimize the time of exposure by carrying food to a safe place rather than by staying in the open and feeding at maximum rate.

3.2.1 Burrowing in Loose Sand

Few animals are able to construct permanent burrows in loose dune sand, but some spiders bind sand grains together with the aid of their silk. One of these is the heteropodid "white lady" (*Leucorchestris arenicola*) of the Namib which inhabits silk-lined burrows 33 ± 13 cm long dug at an angle into the sand. These burrows are excavated by removing sand from the base of a circular depression. The leg coxae and curled pedipalps of the spider bear stiff interlocking setae which are used to push loose sand sideways up to the entrance. Here it is dispersed by flinging it sideways with the brush-like tarsal scopulae. The lower end of the depression is secured by lifting and interweaving loose sand grains with adhesive silk from the spinnerets, and pressing the mixture into the substrate, where it forms a nodule of silk and sand embedded in the surrounding sand. The entrance is closed with a reinforced curtain of silk and sand, whose rim is later severed to form a thin, circular trap-door flush with the dune surface (Henschel 1990b).

Another Namib desert species that burrows into loose dune sand is the eresid *Seothyra henscheli* (Sect. 2.2.1). Although this spider belongs to a different family, it appears to have evolved a similar technique for binding sand grains together with the anterior spinnerets. The striking analogies between the two are a remarkable example of convergent evolution in distantly related taxa, not only in respect of the morphological structures involved, but also with regard to the neurosecretory systems that guide these structures (Peters 1993). For *S. henscheli*, the main outlay in building a new web is the material and labour involved. Relocation is costly, but webs provide the necessary conditions for foraging at most times of the day when prey species are active on the sand surface (Henschel and Lubin 1992).

3.2.2 Sand Swimming

Most of the inhabitants of dune fields do not burrow but, rather, swim in the sand (Sect. 3.2.1). These include fishmoths or Lepismatidae (Watson and Irish 1988), beetles, arachnids and reptiles (Seely 1978). The subject has already been reviewed in this Series at some length (Cloudsley-Thompson 1991) and so will not be discussed again here. Mention may, however, be made of certain Zophosini which have developed a peculiar mode of "sand jumping" in which the hind legs work together instead of alternately (Koch 1955).

Psammoduon (Zodariidae) is an African genus of moderate-sized spiders, some of which, notably *P. deserticola*, have curious habits. They scramble about and burrow through the sand in which they appear almost to swim. Their chief burrowing implements are the specially modified pedipalps. The tarsi of the female pedipalp, especially, bristles with spines and is armed with one or more terminal claws. Permanent burrows are not, however, constructed (Jocqué 1991).

Sand-swimming reptiles use either low amplitude, high frequency, lateral undulations to sink tinto the sand, or else they dive into it head first. The rostrum is pointed and shovel-shaped, the nostrils may be directed upwards and valve-like closure of the eyes, nostrils and mouth enable them to move in loose sand as though it were water. Their bodies are covered with smooth scales which cause little friction, and the legs of lizards may be reduced or even lost (Stebbins 1943). Special morphological and physiological mechanisms facilitate breathing beneath the sand surface (see Cloudsley-Thompson 1991). Sand-swimming mammals include some mole rats (Rodentia), the golden mole (Sect. 3.2) and certain insectivorous moles (Insectivora) of the cool, arid parts of Asia (Wallwork 1982; some 40 species and subspecies of golden moles have been described from southern Africa).

3.3 Rhythmic Activity and Phenology

Not only is the time of activity of a predator influenced by that of its prey (Sect. 2.4) but, conversely, the activity of the prey may be affected by predation – as well as in response to climatic influences. For example, most of the larger Solifugae and scorpions are nocturnal, and the suggestion has been made that these large ground-living arthropods may have become secondarily nocturnal in response to predation (Cloudsley-Thompson 1960b, 1961a). Evidence for this afforded by the fossorial scorpion *Scorpio maurus* of North Africa and the Middle East. This a day-active sit-and-wait predator which mainly ambushes the desert woodlouse *Hemilepistus reaumuri* (Oniscidea), according to Shachak and Brand (1983). Like *H. reaumuri*, and therefore not surprisingly, *S. maurus* shows peaks of activity in the morning and afternoon which, in turn, makes it susceptible to predation by wheatears (*Oenanthe moesta*: Muscicapidae). Of a total of 668 arthropods caught by a single female wheatear to feed her nestlings over a period of 8 days, 9% were *S. maurus* and 34% were *H. reaumuri*. These figures are about equal in terms of biomass (Krapf 1986). Excavation of woodlouse burrows depends upon social cooperation (Sect. 3.9).

It is usually impossible to know whether nocturnal activity has been selected in response to adverse climatic conditions during the day, as a means of reducing predation or, in the case of predators themselves, because their prey is active at night – no doubt several different selective factors are involved simulaneously (Chap. 8). In a recent paper, however, on the evolution of nocturnality in bats, Rydell and Speakman (1995) argue that predation by owls, small hawks, falcons and roller-like birds during the early Eocene was probably a more significant factor in preventing the early bats from becoming diurnal than competition with aerial insectivorous birds. During the Eocene, apart from small Aegialornithidae, there were very few small insectivorous birds, most of the major groups evolving later according to the fossil record.

Seasonal activities among the faunas of arid regions have already been mentioned briefly (Sect. 2.4). Even animals that are active throughout the year and neither aestivate nor hibernate usually show seasonal rhythms of activity and, especially, seasonal reproductive cycles. Seasonal rhythms occur in the diel activity of some species, both of arthropods and of reptiles, which are nocturnal in summer but crepuscular or day-active during the cool season of the year. Annual rhythms such as these are correlated with climatic factors rather than with predation. Nevertheless, the virtual disappearance of a principal prey species throughout much of the year cannot fail to have a detrimental effect on its potential predators.

The distinction between hot desert and cool, moist temperate ecosystems has been defined by Noy-Meir (1973) as a contrast between two paradigms (submodels or modules). The hot desert system can be identified with a "trigger-pulse-reserve" phenomenon, while cool, moist temperate ecosystems conform to a "level-controlling-flows" paradigm. The latter is based on the premise that the energy level in any compartment of an ecosystem is regulated by rates of inflow and outflow, while the pulse-and-reserve paradigm postulates a discontinuous flow of energy through compartments of an ecosystem. According to this model, the trigger (rainfall) initiates a pulse of biological activities. Much of the production is lost throught mortality factors and recycled by detritivores (C. S. Crawford 1991a) (Sect. 7.8), but some is directed into a reserve – seeds and eggs, for example (Wallwork 1982). Some of this reserve is harvested by ants, termites or rodents (Chap. 6) for consumption during the dry season and these, in turn, provide a regular supply of food for predators.

The appearance of some predators is synchronized, by climatic factors, with that of their prey. The inhabitants of desert rainpools – both herbivores and predators – hatch from diapausing eggs when these are wetted. The Australian stumpy-tailed lizard [*Tiliqua rugosa* (= *Trachysaurus rugosus*): Scincidae] feeds on arthropods, snails, carrion, flowers, fruit and berries when such items become available after rain. It is adapted to cope with drought and starvation, when inactive during the dry season, by storing fat in its tail. Effects due to short-term shifts in community composition may be important to desert organisms that have relatively poor powers of dispersal, such as desert plants and arthropods, and which are strongly affected by climatic changes. In general, phenological phenomena in both plants and animals may play some part in the reduction of herbivory and predation but, in arid regions, it is primarily an adaptation to seasonal drought and unpredictable rainfall.

In temperate regions, seasonal changes in the development and physiological state of most living organisms are synchronized by environmental changes, of which photoperiod is by far the most important. Bird migration and seasonal reproduction are well-known examples. In the tropics, and including subtropical deserts, changes in day length are less marked than in temperate regions, and the environmental factor of greatest biological importance is rainfall. Although reproduction is controlled by photoperiod among desert birds and mammals, as it is in temperate regions, it often needs to be triggered by some immediate stimulus from the environment, such as rainfall or the appearance of green vegetation.

Maturation of desert locusts (*Schistocerca gregaria*) occurs in response to terpenoids and other aromatic compounds produced seasonally, under the influence of their circannual clocks, by certain desert shrubs at the time of the annual rains (Carlisle et al. 1965). These compounds are responsible for the characteristic scents of frankincense and myrrh, whose function is to deter herbivorous insects (Sect. 6.2). Not only do desert locusts reproduce when grass is present for their offspring to feed on, but, by the synchronous appearance of the young, predatory enemies are satiated and can only kill a small proportion of them. There is here an interesting interaction between the phenology of desert shrubs and locusts, the food and reproduction of locusts, and also one of their anti-predator devices.

3.4 Crypsis

Buxton (1923) was among the first to draw attention to the fact that, on a broad scale, the inhabitants of deserts tend to be either black, or else buff, sandy, or reddish grey in colour, so that they resemble the background against which they live. The function of black is primarily aposematic (Sect. 3.8), that of desert coloration is one of concealment from enemies. Crypsis has been reviewed by many authors (e.g. Cott 1940; M. Edmunds 1974; Cloudsley-Thompson 1979, 1991; G. Costa 1995 etc.), so only selected examples will be discussed here.

Although Buxton (1923) maintained that the theory of protective coloration could not be invoked as an explanation of the phenomenon of desert coloration, and suggested that it must result from some unknown physiological influence of the environment, it seems to most zoologists today that crypsis is a sufficient explanation (Niethammer 1959; Maclean 1996). The question of the causation of colour is not, however, as simple as it may sound. There is the physiological process itself, as well as its genetical basis involving the action of natural selection, which "causes" a certain gene complex, determining the physiological process, to evolve (Nelson 1973).

In many desert animals, not only does the dorsal surface closely match the colour of the sandy background, but the underside is very pale or even quite white. This "obliterative countershading" counteracts the effect of shadow – the ventral surface appears to be darker than the dorsal even when it is exactly the same colour, because the sunlight comes from above. The contrast in colour between the dorsal and ventral surfaces is especially marked among large day-active desert mammals which are exposed to exceptionally bright sunlight, but the phenomenon is also exhibited by woodlice, centipedes, insects, spiders, lizards, snakes, birds and nocturnal mammals. Furthermore, desert-dwelling animals are not coloured fawn, brown, cream or grey indiscriminately. There is often a very close similarity between an animal and the soil of the particular type of desert in which it is living. Numerous Orthoptera, for example, closely match the colour of the soil in the deserts they inhabit (Isely 1938). Again, many African butterflies show two seasonal colour forms, of which that of the dry season is said to be the one better adapted for concealment. Surprisingly, cryptic patterns on insect wings are almost invariably symmetrical, which may assist predators to detect them. Presumably there is a trade-off between the genetic cost of asymmetry and the extra protection that it would provide.

Apart from tenebrionid and scarabeid beetles, which are mostly black, wasps, ants and certain flies (Sect. 3.8), the majority of desert insects are cryptic and closely match the background colour of the desert in which they live. This is particularly true of grasshoppers (and the solitary phases of locusts), crickets, mantids, termites, caterpillars and some beetles. Desert woodlice, centipedes and arachnids, too, are usually cryptic. Wax deposits on the surface of the integument of the Namib desert tenebrionid *Onymacris rugatipennis* blend with the colour of the sand and provide effective concealment, in addition to reducing water loss (Hadley 1985).

Indeed, most day-active desert animals escape the notice of predators thanks to their cryptic coloration and behaviour. Many jumping spiders (Salticidae), for instance *Menemerus semilimbatus* (Fig. 9), an inhabitant of sand dunes throughout southern Europe and North Africa, are quite inconspicuous unless they move, which they must do in order

Fig. 9. *Menemerus semilimbatus*, a cryptic dune-dwelling spider of the Mediterranean region and Middle East

Fig. 10. *Hersilia caudata*, a cryptic bark spider of Africa and the Middle East. (Cloudsley-Thompson 1984b)

to capture prey. A cryptic animal, therefore, has two solutions to the problem: either it can remain motionless during the hours of daylight and be active at night, or it can move very slowly and stealthily during the day like some spiders, mantids and chameleons (Sect. 2.2). There are, however, some notable exceptions to this generalization, expecially among fast-moving hunting spiders, lizards and small birds. The European wolf spider *Arctosa perita* (Lycosidae), for instance, like *M. semilimbatus*, has a mottled coloration very much like that of the sand on which it lives. When disturbed, it moves very rapidly and then abruptly stops, blending so well with its background that it becomes completely invisible to the human eye, just as insects with flash coloration (Cott 1940) seem to vanish when they become still. This behaviour is probably typical of wolf spiders in general. The bark of trees is a favoured habitat for cryptic spiders, especially in tropical rainforest, but also in deserts. *Hersilia caudata* (Hersiliidae; Fig. 10) is a typical example of a bark spider from the arid lands fringing the Sahara.

Desert amphibians are always cryptic. Tortoises, lizards and snakes are likewise usually desert-coloured except on lava sand, when they are black, or on white sand, where they are extremely pale. Several species of lizards are able to change colour, and this can be correlated both with concealment and thermoregulation. The ability to reflect light is more marked among desert lizards than among species from other regions, suggesting that colour has a thermal influence: but the operation of concealing and thermoregulatory coloration are largely synergistic. When the two produce opposite effects, however, the one possessing the greater survival value is selected. Except under conditions of stress, crypsis is usually the more important (see Cloudsley-Thompson 1991). Both colour change and slow movement enhance crypsis in chameleons. A series of experimental manipulations involving dummy nests and painted ostrich eggs has demonstrated that risk of predation is increased because nests containing conspicuous eggs are more likely to be detected, but the risk has to be taken because darker and more cryptically coloured eggs tend to overheat (Bertram 1992).

Desert coloration predominates among birds, and in no family is it better developed than in the larks (Alaudidae) – typically small, ground-dwelling passerines of arid and semi-arid plains throughout Africa and Eurasia. Numerous authors have commented on desert coloration in birds, with special attention to larks in Arabia, North and Southwestern Africa (for references see Willoughby 1969). Other families of birds showing marked desert coloration include bustards (Otididae), coursers

Fig. 11. Baby ostriches (*Struthio camelus*) on black cracking clay (Sudan). (Cloudsley-Thompson 1965)

(Glareolidae), plovers (Charadriidae), hawks (Accipitridae), weavers (Ploceidae) and sandgrouse (Pteroclidae). The species with desert coloration tend to be sedentary and, in the Namib desert, occur as endemic species and subspecies. Those with generalized cryptic coloration, including young ostriches (*Struthio camelus*: Struthionidae) (Fig. 11), tend to be widely distributed, without forms endemic to the Namib, and with few subspecies.

According to Willoughby (1969), although pale colours may be useful to a bird in conserving water when exposed to hot sunshine, owing to reduced absorption of heat in the visible spectrum of the plumage, this is probably not a significant factor in determining the coloration of desert birds in general. Those few that are conspicuously black or black and white, such as male ostriches, may be so coloured for intraspecific and interspecific social reasons which are more important for the survival of those particular species than is camouflage. Ravens and crows may be conspicuous, but they are formidable birds and presumably extremely unpalatable (Cott 1947).

As Serventy (1971) and Hamiltion (1973) have emphasised, natural selection undoubtedly accounts for the evolution of crypsis. Races of animals living in warm, dry areas are lighter in colour and with a reduction of melanin compared with related forms in more humid regions (Gloger's rule). The physiological and biochemical basis of this is not clear, but its adaptive function is almost invariably crypsis. There can be no other explanation – apart from thermoregulation, which would not apply in the case of nocturnal species.

The plumage of the young of the Saharan sandgrouse *Pterocles senegallus* (Pteroclidae) has two distinctive types of coloration. At the time of hatching they are covered with grey-brown down, but after 4–6 days they change to the sandy yellow colour of the adults. The grey-brown chicks hide among stones of similar colour but, when they become sandy yellow, they crouch on small patches of windblown sand in the lee of bushes and shrubs. Here they dig themselves in until their bodies are half hidden by sand. The young of *P. coronatus*, in contrast, do not appear to undergo such colour change and, at hatching, are similar to their parents (George 1970).

Crouching sandgrouse are not easily seen by jackals and foxes, which prey on the chicks and eggs (Maclean 1968; George 1970). Unless surprised on the nest, sandgrouse usually leave when a predator is 100 m or even further away, whereas the reaction to a non-predatory intruder is to fan the tail suddenly to divert the intruder from the nest. Sandgrouse with young perform elaborate injury-feigning displays to any kind of intruder (for references see Maclean 1976, 1996).

Most desert mammals are exceptionally cryptic. This is especially apparent among rodents in different parts of the world, despite the fact that they are nearly all nocturnal in habit (Harrison 1975). The majority are pallid in colour, their generally sandy or greyish pelage blending harmoniously with the surroundings. This coloration is genetically determined, since specimens bred in captivity far removed from their normal environment continue to reproduce the same coloration for many generations. Dice and Blossom (1937) noted pale, intermediate and dark "substrate races" of *Perognathus intermedius*, *Peromyscus eremicus*, *Neotoma lepida* and *N. albigula* (Cricetidae) occurring on pale, intermediate and dark-coloured rocks in the North American desert. They pointed out that this could not be due to coincidence or to climatic variation.

Hoffmeister (1956) noted the occurrence of dark lava sub-species and pale sand desert races among rodents, while B. H. Baker (1960) found that 14 out of 28 mammal species inhabiting the Durango lava field had more or less darker coats than elsewhere and, significantly, these included all but one of the small ground rodent species with limited home ranges. As Harrison (1975) emphasised, it is no coincidence that the most beautiful sandy-isabellene desert coloration is developed in those Arabian gerbils which are strict psammophiles – such as *Gerbillus cheesmani*, *G. gerbillus* and *G. andersoni* (Muridae). Numerous other examples are cited in the review by Cloudsley-Thompson (1979). The dorcas gazelle (*Gazella dorcas*), the gerenuk (*Lithocranius walleri*) and the addax (*Addax nasomaculatus*) provide excellent examples of desert coloration – pale fawn colours on the dorsal surface with a belly that is almost white. This obliterative countershading (see above) counteracts the effects of shade. Desert carnivores, such as the fennec fox, are also extremely cryptic, but this may be both defensive and offensive (Sect. 2.2).

Eggs and young animals are particularly vulnerable to predatory enemies. Unable to escape by flight, they most frequently resort to concealment if they are to escape unwelcome attention. Even birds that are conspicuous as adults, such as ostriches, frequently have cryptic young (Fig. 11). In contrast, the eggs of birds that nest in holes where they are hidden from view, are invariably white, and the hatchlings are never camouflaged. Some birds construct elaborate nests and shelter for their eggs. African weavers (Ploceidae) build suspended nests, beautifully woven from vegetable fibres. These nests are often attached to the tips of twigs or palm fonds – quite beyond the reach of snakes and other enemies, whose weight would prevent them from reaching the eggs. Such nests are quite conspicuous, especially in the case of social weavers (*Plocepasser* spp.) but, since the eggs they contain cannot be seen, they do not appear to be interesting to would-be predators. Weaver bird nests are frequently sited over water, when this is present, to provide additional security, or very close to the nests of wasps, of larger fierce birds, such as hawks and eagles, or to human habitations.

Birds are not the only animals to conceal their young. Many reptiles bury their eggs in the ground, while the young of small desert mammals are born hidden in burrows. Those of larger, non-fossorial species usually come into the world endowed with cryptic colours and behaviour. Baby rhinoceroses are not cryptic, but their mothers protect them from hyaenas and other predators. Marsupial babies are born in a relatively undeveloped state and at once climb into the safety of their mother's pouch, where they remain safely hidden from the outside world. Every form of deception practised among vertebrates is paralleled in the invertebrate world, and there are many others that are unique.

Both the function and causation of desert coloration have engendered much dispute. For instance, Buxton (1923) found it inconsistent that both predators and prey, running and burrowing mammals and birds were all camouflaged. "The desert-coloured owl chasing a desert-coloured bat over desert-coloured soil under a jaundiced moon", was how be expressed it! Escape, he claimed, depends mainly upon speed, and birds of prey can seldom catch a sand grouse, a lark or a bustard in flight. Yet these are among the most cryptic of desert birds. In fact, however, although a sandgrouse may escape from a hawk by flight, it does not always do so. Therefore it is better, when on the ground, for it not to be seen at all and not to have to take to the air. As Cott (1940) emphasised, even small advantages may have selective significance over thousands of years. The main function of desert coloration is really no longer in doubt. Other possible functions are outlined by Dorst (1974). The matter was discussed most ably by Nelson (1973) and, more recently, by G. Costa (1995). The interaction between predators and their prey is dynamic, and constantly being modified by natural selection. There are, however, limitations to the perfection of crypsis because the adaptations for concealment must inevitably conflict with other essential activities.

3.5 Protective Resemblance and Disguise

The selective process resulting in protective resemblance, in which an animal is disguised as a stick, bark, a stone or some other inanimate object that is of no significance or interest to potential predators (Cott 1940) is probably the same as that which produces mimicry. Some animals that are cryptic from a distance display protective resemblance when seen close up (Vane-Wright 1980, 1981). Starrett (1993) uses the term adaptive resemblance to unify the concepts of mimicry and crypsis.

In arid country, naturally enough, many of the most common inedible objects are stones, and stone mimicry, or protective resemblance to stones, is not uncommon. Crickets and grasshoppers (e.g. *Batrochornis perloides*), having fairly solid bodies, not infrequently look like stones, while mantids, with long thin bodies, more often resemble twigs (see Cloudsley-Thompson 1991, 1995). Mantids of the family Eremiaphilidae, e.g. *Eremiaphila* spp., however, live on the ground and mimic stones. They are not uncommon throughout North Africa and the Great Palaearctic desert. Spiders of the genus *Caerostris* (Araneidae) leave their webs during the day, and rest on spiny twigs, where they look like broken spines (P. Ward 1979). A fine example of protective resemblance is afforded by the round-tailed horned lizard *Phrynosoma modestum* (Iguanidae). From a distance, this small species matches its background but, close up, it could easily be mistaken for a stone (Sherbrooke and Montanucci 1988).

3.6 Mimicry

True mimicry is of two kinds – batesian and mullerian. In classical batesian mimicry (first described by H. W. Bates in 1862), a harmless and vulnerable species resembles an unpalatable or dangerous model that is ignored by potential predators. Harmless hover-flies (Syrphidae) are often cited as mimics of poisonous wasps, bees and ants: bombardier beetles are mimicked by grasshoppers, ants by spiders, poisonous moths by harmless species and so on. Batesian mimicry is similar to disguise or protective resemblance (Sect. 3.5) in that it is a means of deceiving enemies. Whereas anachoresis and crypsis can prevent a predator from detecting its prey, however, disguise, mimicry, and aposematism or warning coloration (Sect. 3.8) merely ensure that the prey is either not recognised as anything edible or else seen as something positively to be avoided. A noxious species benefits from warning coloration or sounds, because some of its members are sacrificed, or have to defend themselves, in teaching would-be predators to avoid them. Therefore, if one or more aposematic species mimic one another, the losses in teaching enemies not to attack them are shared and proportionately reduced.

The mimicry of one distasteful model by another distasteful species is known as mullerian mimicry (first described by F. Müller in 1879; for a brief historical account, see Cloudsley-Thompson 1980). Both batesian and mullerian mimicry are common among desert arthropods. Aposematic animals are almost invariably black in the desert (Sect. 3.8). Black desert bees and wasps could, therefore, be models for black Bombyliidae. Alternatively, it is by no means improbable that desert beeflies gain from being conspicuous and drawing attention to themselves. They are speedy, agile insects scarcely worth chasing. It benefits their potential predators to realise this fact, and they and their mimics gain by not having to expend energy in escaping. This phenomenon is known as "speed" mimicry (see Cloudsley-Thompson 1991). The unique occurrence of a complex of true scarabs with orange elytra in the Namib desert has been described by Holm and Kirsten (1979), who explain the adaptation in terms of speed mimicry. Alate species have orange elytra and are thereby distinguished from the predominantly black, wingless species. It is argued that one wingless species, *Pachysoma denticolle*, is a batesian mimic of *Scarabaeus rubripennis*, while the latter and four other alate species are mullerian mimics of one another.

It has previously been suggested that the great flexibility between the prosoma and opisthosoma of Solifugae, which enables the opisthosoma to be held in an almost vertical position, may be a form of scorpion mimicry affording some protection from predators (Cloudsley-Thompson 1958). Apart from their relatively enormous jaws which could, presumably, inflict a sharp bite, Solifugae are relatively defenceless against larger vertebrate enemies.

Many jumping spiders (Salticidae) are batesian mimics of ants (Formicidae) or, to a lesser extent, of Mutillidae. The males of *Cosmophasis nigrocyanea*, for instance, resemble ants, while the females look like mutilled wasps. Another instance of ant mimicry is provided by *Seothyra henscheli* (Sects. 2.2, 2.2.1 and 3.2.1). Ant-mimicking spiders have long, slender legs and, in some species, the pedical is constricted so that it resembles the waist of an ant. Alternatively, pale bands across the body may contribute to the deception. Several families of spiders are involved in ant-mimicry, but the phenomenon is most marked among Corinnidae and Salticidae, especially *Myrmarachne* spp. Morphological and behavioral mimicry of ants has been reviewed by McIver and Stonedahl (1993).

Most desert lizards possess cryptic coloration, but juveniles of the species *Eremias lugubris* (Lacertidae) are batesian mimics of carabid beetles of the genus *Thermophilum* (= *Anthia*) which spray an acidic, pungent fluid when molested (Huey and Pianka 1977), while the gecko *Coleonyx variegatus* mimics large scorpions of the genus *Hadrurus* (Parker and Pianka 1974).

Mimicry is not common among birds, and some of the reported cases may be due to coincidence rather than to adaptation, especially when the two species concerned belong to the same taxonomic group. The cuckoo (*Cuculus conorus*), however, shows a resemblance to the sparrowhawk (*Accipiter niser*) which extends to the immature plumage. Again, the black drongo cuckoo (*Surniculus lugubris*) resembles the pugnacious black drongo (*Dicrurus macrocercus*) which frequently attacks and pursues certain large birds such as crows and hawks. Examples from true desert species have not often been observed, but the zone-tailed hawk (*Buteo albonotatus*) differs from related species and closely resembles the American turkey vulture (*Cathartes aura*) in colour, shape and manner of soaring. The hawk often soars with vultures and dives on its prey from groups of vultures. This is possibly a case of aggressive mimicry, because prey animals become accustomed to the passage of inoffensive vultures (Willis 1963). At the same time, vultures are extremely distasteful (Cott 1947), whereas hawks are probably not.

Very few examples of mimicry are known to occur among mammals – even fewer than in birds. This may partly be due to the fact that mammals are, in general, more difficult animals to study in their natural environments. Many of them are nocturnal, some are extremely active and tend to travel considerable distances, and most are shy and difficult to observe. At the same time, mammals tend to be long-lived and reproduce rather slowly, so that the evolution of specific mimicry must take a good deal longer than it does in lower animals.

It has been suggested, however, that the small and inoffensive aard-wolf (*Proteles cristatus*) may have developed its close visual resemblance to the larger striped hyaena (*Hyaena hyaena*) for protection against predators that hunt by sight (Gingerich 1975). These predators are mainly leopards, but the lion would certainly eat an aard-wolf if it got a chance (see Cloudsley-Thompson 1980). On the other hand, Goodhart (1975) argued against the mimicry hypothesis. He pointed out that, in southern Africa, the striped hyaena is replaced by the brown hyaena (*H. brunnea*) which does not look much like an aard-wolf. It is more likely, therefore, that the resemblance is due simply to close phylogenetic relationship, both species having retained the cryptic coloration of a common ancestor, together with its erectile dorsal crest, used for threat display as in many other mammals. Again, young cheetahs (*Acinonyx jubatus*) appear to mimic ratels or honey-badgers (*Mellivora capensis*: Musteilidae), which are aggressive and pugnaceous, defending themselves by emitting an offensive, suffocating odour from the anal glands. Both ratels and cheetah kittens show reversed countershading which enhances their conspicuousness (Sect. 3.7) and makes them look alike.

Mullerian mimicry is benefical not only to potential prey which sacrifice fewer of their numbers in teaching potential predators to avoid them, but also to the predators themselves. It reduces the educational burden on them, allowing them to learn to avoid two or more unpalatable species with a sampling effort smaller than would be needed if the prey were not mullerian mimics. Whether a predator gains from such mimicry is crucial factor in predator-prey co-evolution, since a predator that gains from mimicry between two or more potential prey species will not be likely to put much effort into discriminating between these species (Fisher 1930). In contrast, a predator whose fitness is lowered as a result of mullerian mimicry (because one species of the complex is less unpalatable than the others) is likely to invest more effort in scrutinizing the various species which form a spectrum of unpalatability.

Speed (1993), however, has argued that mimicry between unpalatable prey species does not automatically confer a benefit on their predators. Although there are clearly benefits in certain circumstances (e.g. reduced educational burden), there are also potential costs, too (losses of life-saving food). It is possible for either of these to outweigh the other when assessed throughout the whole life of the predator, and mimicry by unpalatable prey will not necessarily aid the predator. Lane and Rothschild (1965) cite a possible instance of sound mimicry between bumble bees (*Bombus* spp.) and the burying beetle *Necrophorus investigator*, whose stridulations and behaviour are remarkably similar.

An important distinction should be drawn between the evolution of warning coloration, and of the unpalatabilty that it advertises. The conditions favouring unpalability themselves affect the conditions that engender aposematism. Both individual and kin selection must be involved, because related individuals are most likely to share the gene for unpalatability. At the same time, once warning coloration has evolved, predators may use specific colour cues to recognize unpalatable individuals so that the individual benefits from being unpalatable are reduced. It may now be possible for automimicry to evolve to the point at which there is a stable equilibrium between more palatable and less palatable individuals when the cost of being unpalatable balances the increased personal risk associated with not being so (see discussion in Guilford 1990).

3.7 Integument, Scales and Armour

Unlike venoms and defensive fluids (Sect. 4.6), which are secondary defences, invoked only in the presence of a predator, the animals that possess them carry their primary defenses throughout their lives. It might be argued that spines and urticating hairs (Sect. 4.5) are also present throughout the lives of their possessors, but they are usually erected or discharged only under attack. The arthropodan cuticle or integument is often hardened and has a decidely defensive function. Many desert beetles, especially of the families Tenebrionidae, Histeridae, Scarabeidae, Curculionidae and Carabidae, have both primary defences (hardened cuticles) and secondary defences (distastefulness, Sect. 4.6). They can be kept in a vivarium for several months with large scorpions, such as *Androctonus australis* (Buthidae) before they are attacked and eaten. There are few, if any, predators that will eat large tenebrionids unless stressed by hunger. The only exceptions are Solifugae: female *Galeodes granti* (Galeodidae), for example, readily devour hard desert beetles such as *Pimelia grandis* (Tenebrionidae), crushing their integument to produce a noise like the cracking of nuts at a Christmas party. Males are more hesitant, and usually respond by threat movements, followed by flight (Cloudsley-Thompson 1961b). *Galeodes* spp. in West Africa also devour large black millipedes (M. Edmunds, pers. comm.).

Many beetles, including numerous desert species (Lewis 1964), have a very compact and hard exoskeleton (as in most weevils) or a smoothly rounded and almost ungraspable upper surface (as in many Chrysomelinae and Eumolpinae; Crowson 1981).

Some desert scorpions have a rather hard, thick, integument on parts, if not the whole of their bodies – for example, the claws of scorpionid scorpions and the tails of *Androctonus* spp. – but the function of this is not always anti-predatory. The claws of *Scorpio maurus* (Scorpionidae), for instance, are important tools for digging, while *Androctonus* is one of the most venomous of all genera of scorpions and its poison is readily employed for defensive purposes (Sect. 4.6.1). Nevertheless, in some scorpion genera, including *Androctonus*, *Orthochirus* and *Apistobuthus*, in which the tail is robust and spiny, the second basal segment is enlarged to form a disc which is strongly reinforced ventrally with heavy keels to form a shield. When the tail is curled up, the formidable sting points forward, protecting the front, while the thick shield guards the rear of the tail. This would be a useful adaptation in the narrow burrow of a snake or rodent if the scorpion were attacked from behind. At other times, the armoured tail protects the delicate dorsal surface of the opisthosoma, which must be supple to allow the scorpion to ingest a large meal.

Most of a desert amphibian's life may be spent in aestivation. Spadefoot toads (*Scaphiopus couchii*), *S. hammondii* and the African bullfrog (*Pyxicephalus adspersus*: Ranidae), for example, commonly aestivate for 7–10 months per year (Sect. 5.3) and are protected by their coccoons, not only from desiccation, but also from predatory enemies and parasites.

The scales of reptiles are protective as, to a much greater extent, are the shells of tortoises and terrapins (Cloudsley-Thompson 1994). Reptilian scales are composed of keratin, which is continually being rubbed away and renewed, mainly from the epidermal tissues beneath. Among lizards and snakes, it is sloughed at intervals, sometimes several times a year. In some desert lizards, the dermal parts of the scales contain small osteoderms which make the skin exceedingly tough. In the head region, these plates may be attached to the underlying bones.

Girdle-tailed lizards or zonures (Cordylidae) of southern Africa have developed a remarkable armour of spines (Fig. 12). Not only does this armour make them difficult for their predators to swallow, but it enables them to wedge themselves into crevices whence it is extremely difficult to extricate them. Spines are also accentuated in spiny lizards (*Sceloporus* spp.: Iguanidae), *Ctenophorus* (Fig. 12) and *Calotes* spp. (Agamidae), as well as in spiny-tailed lizards (*Uromastyx* spp.) in which the tail is a formidable defensive weapon. Indeed, second to teeth, tails are possibly the most effective weapons of defence amongst reptiles and are employed with special effect by monitors (Varanidae), large iguanids

Fig. 12. *Above* Girdle-tailed lizard (*Cordylus giganteus*) (Southern Africa); *below* crested dragon (*Ctenophorus cristatus*) showing spiny scales (Australia)

and agamids. Indeed, the Iguanidae and Agamidae show remarkable parallels in form, structure and habits (Pianka 1985, 1986; Cloudsley-Thompson 1994; Sec. 1.2).

The chelonian shell consists of a thin layer of keratin plates or laminae, and an inner layer of bony plates. The laminae do not coincide with the bony plates but overlap them, thereby adding strength to the structure. The upper carapace is joined on either side to the ventral plastron by a bridge formed by an upgrowth of the latter. In the North American box tortoises (*Terrapene* spp.: Emydidae) there is a hinge across the middle of the plastron, while in African tortoises of the genus *Kinixys* (Testudinidae), this carapace is hinged, only allowing the rear of the shell to close. Mechanisms for closing off one or both openings of the shell must have evolved independently on several occasions in tortoises and terrapins. It is obvious, therefore, that some predators must be able to kill a retracted chelonian, but only if the openings to its shell are not closed. In the large African *Geochelone sulcata* the front opening to the shell is closed by the heavily armoured fore-legs, the rear one by the tail and back legs (Cloudsley-Thompson 1994).

3.8 Aposematic Coloration

Aposematic or warning colours are characteristic of distasteful, venomous, and otherwise formidable animals. Primary anti-predator devices in themselves, they depend for their success upon the presence of secondary devices (Chap. 4) unless they occur in mimics of the distasteful or dangerous species. The conspicuousness they confer renders their bearers readily recognisable, so that predatory enemies soon learn to avoid them. Some predators may have an innate aversion to certain aposematic signals. Mullerian mimicry (Sect. 3.6), as we have seen, reduces losses involved in teaching enemies not to attack, and may also benefit potential predators by reducing the educational burden placed upon them.

Aposematic desert animals are almost invariably black. For a yellow wasp to become conspicuous it must acquire black markings. On a yellowish background of desert sand, black is far more conspicuous than black and yellows – reds, yellows and browns do not show up well. At the same time, an element of mullerian mimicry (Sect. 3.6) may also be involved in the almost universal acquisition of black (Cloudsley-Thompson 1979). Aposematic birds, like many other animals, tend to be distasteful (Cott 1947).

Some conspicuous animals are inedible, others may be batesian mimics of distasteful or dangerous species. The important fact is that they should be distinctive and easily recognised. This could explain why some distasteful models are not especially conspicuous although they are undoubtedly distinctive. Examples are afforded by various Lepidoptera (Papilionidae, Danaidae, Heliconiidae) and Neuroptera (Ithoniidae), some of which contain alkaloids and other poisonous chemicals, while others are batesian mimics of these. Many such toxic species, although far from inconspicuous – at least when disturbed – are by no means as conspicuous as venomous black and yellow wasps. At the same time, they and their mimics may possess aposematic scents.

Aposematism is not restricted to colour alone. Warning scents, sounds and deimatic displays are also aposematic. They will be discussed below (Sect. 4.7) since they are usually secondary anti-predator devices, only displayed in the presence of an enemy, whereas warning colours are present at all times.

3.9 Communal Behaviour

There are both advantages and disadvantages to group living. The various benefits and costs differ with the habitats of the different species. The former include increased efficiency in locating and capturing food, in reproduction, as well as in detecting and avoiding predators: costs may involve increased competition, disease transmission and conspicuousness to predators. (Some of these points will be discussed further in Sect. 7.1.)

Social behaviour in desert arthropods has been reviewed comparatively recently (Cloudsley-Thompson 1989, 1991) and will not be discussed in detail here. For survival and reproduction, the desert woodlouse *Hemilepistus reaumuri* is dependent upon its burrows, which have to be defended continuously against intraspecific and, to a lesser extent, interspecific competitors. This requirement is met by the cooperation of individuals within the framework of a complex system of social behaviour within strictly closed family communities (Linsenmair 1984). While they are down their burrows, desert woodlice enjoy an equable microclimate and are secure from their main predators, to which they are exposed when they emerge (Sect. 3.3).

Termites and ants are successful insects of desert regions, to which they are pre-adapted in that they are social and construct underground nests or, in the case of some termites, almost impregnable mounds. Even gregarious, but non-social, animals benefit from group life. A large swarm of locusts, for instance, contains so many individuals that the proportion lost to predators is almost negligible. Hawkmoths (Sphingidae) of numerous species, and field crickets (*Gryllus bimaculatus*: Gryllidae) may become extremely numerous in the Sahel savanna after the rains have ended in September each year. Their numbers are so great that predatory animals must soon become satiated. Social spiders clearly benefit from their mutual cooperation (Cloudsley-Thompson 1991). *Stegodyphus* spp. (Eresidae) build their nests in acacia trees (Fig. 13). They are strictly nocturnal and no signs of activitiy are apparent during the day. As darkness falls, however, the inhabitants emerge from

Fig. 13. Nest of social spiders (*Stegodyphus mimosarum*) (Namib desert)

the entrances to their holes and attack insects trapped by the silk of their communal webs. In contrast to insect societies, colonies of social spiders show no caste system or division of labour. Even outsiders from other colonies of the same species are tolerated, and the nests are shared with numerous commensals (Sect. 5.6).

Reptiles are not particularly social, but several species show parental care (Shine 1980) and their distribution is often clumped (Sect. 7.1.1) – which may afford some protection. When flocks of small birds sight a peregrine falcon, they bunch tightly together, a specific response not employed against other hunters. Falcons are much less likely to attack a tight flock than single birds – presumably because they might harm themselves – and tend to select stragglers. Thus, even a predator as efficient as a falcon may have its efficiency impaired by appropriate behavioral responses of the prey (Pianka 1978).

Many bird species form large flocks and thereby add to their safety. In particular, quelea and sandgrouse deserve mention in this context. The red-billed dioch (*Quelea quelea*: Ploceidae; Sect. 7.1.1) is completely colonial in habit and often found in concentrations of over a million birds – in flight they can be mistaken for locusts! Its main habitat is in areas south of the Sahara with less than 75 mm rain per year, but migrations of over 1200 km have been established by ringing.

The anti-predator behaviour of sandgrouse is especially well developed in their drinking patterns. According to Maclean (1976), three basic patterns exist: some species, such as *Pterocles senegallus* and *P. namaqua*, usually land some distance from a waterhole, coming in to drink only when the coast is clear. Other morning-drinking species (e.g. *P. burchelli* and *P. decoratus*) land right at the water's edge or even in the water. The third pattern is that of *P. lichtensteini*, *P. quadricintus*, and *P. bicinctus*, which drink at dusk. Arriving at the water only in the late evening or after nightfall, they may land at the water or away from it according to circumstances. Once in the water, the birds drink extremely rapidly and then take off again within a few seconds, their wings twinkling in the sunlight.

P. senegallus exhibits an interesting anti-predator manoeuvre at the waterhole. When herds of domestic animals are drinking, with their herders in attendance, the sandgrouse gather 200–1000 m away. The birds on the ground may not be able to see the waterhole, in which case, at irregular intervals, a single bird will take off and fly unusually high (50–80 m) over the waterhole, very slowly and several times. The slow flight is achieved by rapid, shallow wing beats and accompanied by a characteristic call. The bird then returns to the flock and, if the water is free, all the birds then fly to it (George 1970). These "sentry flights" have not been recorded in *P. coronatus* and, although non-breeding *P. senegallus* may wait for up to 10 h, breeding birds cannot be kept from the water for more than an hour, after which they will drink even in the presence of men and domestic stock.

The main predators of sandgrouse at waterholes are raptors of the genus *Falco* (Maclean 1968; George 1969; Willoughby 1969). Moroccan populations of *P. senegallus* are always accompanied by falcons in the proportion of roughly one to every 100 sandgrouse (George 1970). The high selection pressure by predation is especially evident among Saharan sandgrouse populations, whose density varies with the food supply: if this is inadequate, the birds do not breed, whereas the population density of the falcons remains more or less constant because they can always fall back on migrating birds during their own breeding season and on sandgrouse at other times of the year. In open desert habitats, crouching flat is an excellent form of defence, whose value is enhanced by the formation of nesting and roosting scrapes into which the bodies of the sandgrouse merge effectively. Even so, falcons have learnt to fly low over the ground and pick up the silhouettes of crouching sandgrouse (Maclean 1976). This method of hunting cannot be as effective as the more usual one of flying into or over a flock of sandgrouse and catching the odd bird that takes off in fright (Willougby 1969; George 1970), which accounts for the well-developed sandgrouse behaviour of keeping watch at waterholes, and of very rapid drinking.

Colonial breeding is undoubtedly the main defence of queleas (see above). Colonies provide early warning of the approach of a predator: although no alarm signal is given, the nest is not defended and predators are not subjected to communal mobbing (Sect. 7.1.2), as in weaver birds. The main defence of the quelea is to breed in huge, dense, highly synchronized colonies in thorny trees (Jones 1989; Thiollay 1989). The adaptive significance of coloniality in birds has been reviewed by Orians (1971) and by Wittenberger and Hunt (1985; see discussion in G. Costa 1995). Communal behaviour in mammals has been discussed by many authors including Ewer (1968, 1973), Kruuk (1972), Schaller (1972), Sherman et al. (1991) etc.

3.10 Vigilance

Animals that are not cryptic, distasteful, or protected by armour, spines, or weapons of various kinds, need to be alert if they are not to become the victims of predation. In social species (Sect. 7.1.2), certain individuals may keep watch whilst other members of the group forage for food. This is seen in the matriarchal groups of dwarf mongooses (*Helogale parvula*), in which subordinate males regularly alternate in vigilance behaviour, emitting a "song" (Rood 1983, 1990). This keeps other members of the group informed when an enemy approaches - behaviour which is paralleled by that of meerkats. Spring hares (*Pedetes caffer*: Pedetidae) of the Kalahari forage in groups, although they have separate holes.

Increased vigilance results in reduced intake, and there is always a trade-off between foraging and danger. Predators that stalk their prey (Sect. 2.1) rely on surprise. They are seldom successful if the prey is alerted to their presence before the final attack. Vigilance is costly, however, because it interferes with foraging and other activities. Members of groups enjoy an increased probability of detecting predators, and of increasing the distance at which detection occurs without increasing their own level of vigilance because, in a group, there are more individuals on the lookout at any one time (Caro and FitzGibbon 1992). For example, larger groups of Thomson's gazelle (*Gazella thomsoni*) detect approaching cheetahs at greater distances than do smaller groups. Other mammals that benefit from the vigilance of conspecifics include dwarf mongooses, meerkats and ground squirrels (Fig. 14).

Furthermore, when Thomson's gazelles join Grant's gazelles (*G. granti*) to form mixed-species groups, several advantages are obtained.

Fig. 14. Vigilance postures. *Left* Meerkat (*Suricata suricatta*); *right* ground squirrel (*Xerus inaurus*) (Kalahari desert)

Not only is predator detection improved, but cheetahs tend to avoid hunting larger groups. Associating with Grant's gazelles does not enable Thomson's gazelles to spend less time being vigilant than does associating with the same number of conspecifics, but Grant's gazelles benefit more because cheetahs tend to select the smaller Thomson's gazelle when attacking mixed-species groups (FitzGibbon 1990). Oxpeckers (*Buphagus* spp. Sturnidae) do not merely benefit their temporary hosts by removing ticks – although they take blood at the same time – but, by flying round with raucous hissing notes, they warn them of approaching danger.

G. Costa (1995) discusses the alarm signals exchanged between conspecific desert animals, and Maclean (1996) refers to the vigilance of desert birds whilst drinking. Psychological studies have established that the central nervous system cannot sustain vigilance for an extended period of time, and there is a gradual decrement in the ability to detect both hidden predators and cryptic food items. The optimal proportion and optimal length of time devoted to foraging should theoretically be decreasing functions of the rate of vigilance decrement. Because the rate of decrement is higher for more difficult tasks, both the time spent on foraging and the duration of each foraging episode should be smaller for more demanding foraging activities (Dukas and Clark 1995; for further discussion of the vigilance of animals in groups, see Sect. 7.1.3).

The primary anti-predator devices of desert animals are less diverse and numerous than those of animals in the faunas of mesic tropical and subtropical areas but, where they are appropriate, they have been exploited to an extreme engendered by the openness and relative lack of vegetation.

4 Secondary Anti-Predator Devices

Secondary anti-predator devices are invoked only in the presence of an enemy. The simplest response of most prey animals is to fly or run away, while burrowing forms retreat as fast as they can into their shelters or retreats. This may not always be possible, however, and then self-defence with whatever weapons are available may become necessary. Some animals, such as tortoises or porcupines, are so well endowed with defences that they stand their ground and make no attempt whatever to escape. Others, equipped with formidable offensive weapons – especially teeth – readily employ them for defence.

4.1 Flight and Escape

Only the largest of desert animals, such as ostriches, antelopes and camels, spend all their time exposed in the open. The vast majority are small enough to inhabit burrows and retreats, to which they return when threatened by a predator (see review by G. Costa 1995). They may even possess adaptations that maintain them in security. For instance, denticles on the tails of scorpions help to anchor them in their burrows so that they are not pulled out by predators or large insect prey. Certain insects, such as grasshoppers and mantids, which do not flee to shelter, may enhance their concealment by the display of flash coloration (Cott 1940). Although all visible portions of the resting animals are obliterated by the cryptic coloration which covers their exposed surfaces, an advertising colour scheme is disposed on parts of the body that are normally concealed. This becomes immediately conspicuous when the animals move, revealing a brilliant flash of scarlet, yellow or blue. The hind wings of desert grasshoppers, for instance, are often red or bright blue. The disappearance of this colour the moment that the insects land can be most confusing. Many butterflies illustrate the same principle. The upper surfaces of both wings are brightly coloured but, when the insects settle with the wings folded together above the back, only the cryptic undersurfaces are visible.

A few spiders, including some Thomisidae, are able rapidly to change colour, and this may also be a form of flash coloration. Some frogs and toads, too, have bright colours on the inner surfaces of their hind legs: these are visible only when they jump. Sound, or acoustic flash behaviour, is exhibited by certain grasshoppers which make a noise whilst they are flying, but are silent on the ground. The erratic or protean flight of butterflies and moths makes their capture by birds and bats very much more difficult.

Cartwheeling, with the legs curled, is an unusual mode of escape among desert dunes and recorded only in spiders. *Carparachne aureoflava* (Heteropodidae) uses little energy while rolling down the slope of a dune to reach speeds of 0.5–1.5 ms^{-1} when escaping from pompilid wasps etc. At 10–14 rotations per s, the spider's outline becomes blurred to a vertebrate eye and no longer presents the normal stimulus to its adversary (Henschel 1990a). Some Salticidae of the Namib desert also cartwheel when disturbed.

Most desert lizards are extremely quick and difficult to catch. They dart for cover, which may be a hole in the ground or a rock crevice, with surprising speed – often changing direction quite unexpectedly as they do so. Refuges are not restricted to burrows or rock fissures. As anyone who has tried to catch reptiles in central Australia will know, porcupine grasses (spinifex; Sect. 6.3) provide an almost inaccessible retreat for lizards and snakes.

Lizards that forage widely and rely upon rapid flight to escape predation are usually streamlined and have a low ratio of clutch to body weight. A large clutch of eggs would hamper such species in running, and thus reduce their ability to escape: it would also reduce their foraging efficiency. Cryptic, sit-and-wait foragers, on the other hand, tend to be short and robust, and to produce relatively large clutch masses. Since they do not rely upon speed, either for foraging or for escaping, they can afford to bear heavy clutches. Thus, relative clutch mass, body shape and foraging and escape tactics are all interrelated in lizards (see Cloudsley-Thompson 1994). Not surprisingly, species of lizards with larger body size tend to have greater stamina (Garland 1994).

Two widespread specialized locomotory escape mechanisms are running on soft sand, which is often facilitated by toe fringes (Luke 1986), and climbing relatively smooth, vertical surfaces with the aid of highly evolved morphological adaptations of the digits. These adaptations are found in geckoes, some iguanids, and a few species of skinks (Green 1988).

Desert dunes are inhabited by a distinctive fauna of lizards and snakes which often make their escape by diving head first into the sand (Arnold 1995). Sand swimmers of the Namib desert include *Meroles anchietae* and *M. cuneirostris* (Lacertidae) and the sidewinding viper *Bitis peringuei*. In North Africa and the Middle East *Sphenops sepsoides* and *Scincus* spp. (Scincidae) are typical examples. *Uma* spp. (Iguanidae) and *Chionactis* spp. (Colubridae) of the Mojave desert and the sidewinder *Cerastes cerastes* of the Sahara use low-amplitude, high-frequency lateral undulations to disappear into the substrate (for references see Cloudsley-Thompson 1991, 1994). Many lizards that live in sandy places are extremely speedy and vigilant. Not infrequently, the toes of the feet are broadened by fringers of elongated scales which enable them to run quickly across soft sand.

Birds of open habitats, such as deserts, where there is little cover in which to hide, may deter predators by cooperative manoeuvres whilst in flight (Sect. 7.1.1). Ground foragers are particularly vulnerable to predators and rapidly retreat to shelter or take flight when danger threatens. Those that glean nearby seeds and invertebrate prey are probably less aware of enemies than are sit-and-wait foragers, themselves looking for food several meters away: the latter take more active prey and may have greater mobility, which can be useful when escaping from their own predators (Ford 1989).

In flight, Egyptian geese (*Alopochen aegyptiaca*) and ruddy sheld duck (*Tadorna ferruginea*) (Anatidae) are magnificent birds with broad pinions of black, white and bronze green. When they alight, however, and close their wings, the black and white colours suddenly disappear and the sandy-coloured bodies become practically invisible. Many other birds, too, show sudden flashes of colour, such as white rumps or tail feathers, when they fly. It is possible not only that this is flash coloration but, in addition, it may serve the function of warning other birds of the presence of a predator nearby. This would explain why many species in the tropics which form mixed flocks share the same colours on the rump or under tail coverts (Edmunds 1974).

Small desert mammals are either burrowers or sand digging forms such as jerboas, kangaroo-rats and ground squirrels (Fig. 15) that flee quickly to safety when discovered by a predator. Even insectivorous Microchiroptera inhabit holes, rock crevices or caves during the day. When they emerge at dusk, they become vulnerable to kites (*Milvus migrans*) and hawks. Larger desert mammals, such as gazelles and antelope, either escape at high speeed or defend themselves with horns and hooves (Sect. 4.8). Addax, which have less formidable defences than oryx (*Oryx* spp.), inhabit such arid, inhospitable wastes that they are seldom exposed to attack by large predators. Moreover, their hooves are considerably enlarged as an adaptation to walking on soft, sandy soil (Sect. 8.1.4).

Escape and flight may thus be of various types. If the prey is faster than its predator, it will often escape by direct flight. Erratic and unpredictable protean movements may be successful when the prey is less speedy but able to flee into a burrow or retreat where its enemy cannot follow it. Protean movements, involving zigzag turns and twists, make the escape route difficult to predict, and thereby upset the pursuer, which may be unable either to change its course equally quickly or benefit from learning from experience what its quarry is likely to do

Fig. 15. American ground squirrel (*Citellus* sp.) beside its burrow (Sonoran desert)

(Driver and Humphries 1988). A third technique is for the prey to flee for a short distance, often displaying flash coloration, and then to rest, motionless and cryptic, so that it cannot be observed (Edmunds 1974). Although escape movements are very important in all environments, they are particularly important in desert and arid regions where plant cover is scarce or non-existent and it may be necessary to forage some distance from the burrow or retreat.

When they see a predator, such as a cheetah or leopard, Thomson's gazelle will follow it – perhaps so that they can keep an eye on their dangerous enemy. They react mainly to optical stimuli provided by the silhouettes or movements of the predator. "Stotting" is often their first response (Walther 1969). Stotting – running and jumping in a jerky manner and exposing the white rump patch (Fig. 16) – is characteristic of gazelles. It may be an indication to the predator that it has been observed by its speedy prey which, in the springbok (*Antidorcas marsupialis*), from which it gets its name, is the habit of jumping into the air with the attitude of a bucking horse. The head is lowered almost to the feet, the legs hang fully extended with the hooves almost touching, which arches the back. "Pronking", as this is called, is probably an exaggerated form of stotting, and may well have a similar function. It enables the buck to see in all directions for a considerable distance. Even kids, only a

Fig. 16. Stotting gait of a Thomson's gazelle (*Gazella thomsoni*) adopted expecially when approached by a predator

day or two old, do it. At the same time, it is an intraspecific signal and affords warning to other members of the herd, as it is so conspicuous.

Prey species, such as Thomson's gazelle, are attacked by numerous predators including hyaenas, jackals, baboons, martial eagles (*Polemaetus bellicosus*) and vultures, from which they can escape only by fleeing. The flight distance is greater the more dangerous the predator, and it is also greater if more than one carnivore is present (Kruuk 1972). The flight distance of gazelles is thus less for jackals than for hyaenas, increasing for lion, leopard and cheetah, whilst wild dogs are avoided from a distance of a kilometer or more (Schaller 1972). Prey species are usually well adapted to their predators, however, and populations remain in balance in arid regions even if they are spatially more extended than in more humid regions.

4.2 Thanatosis

Animals that are unable to evade their predators sometimes appear to "give up" and become motionless, as though they were dead. Death-feigning, or thanatosis, can inhibit attack because the prey fails to release a killing response in its predator. The best-known example of this is seen in the American opossum (*Didelphis marsupialis*: Didelphidae) from which we get the expression "to play possum" (Cloudsley-Thompson 1980). Many insects of various kinds also become inert when they are attacked, either with the limbs extended, or withdrawn close to the body so that they cannot be bitten off.

The first choice with which a spider is presented when a potential prey item becomes ensnared in its web is whether or not to attack it. By means of vibrational stimuli, the web provides it with information about the size, activity and location of the prey. If the insect prey is too large or too active, the spider may retreat. If it is very small or inactive, the spider may ignore it. When trapped in a web, some insects alternate long periods of inactivity with brief, violent struggles to free themselves. This strategy may depend upon the fact that, in many cases, only prolonged, low-amplitude vibrational stimuli provoke attack. It can be observed in Bibionidae, Rhagionidae, Coccinellidae, Chrysomelidae, Symphyta, lepidopteran larvae and Opiliones (Nentwig 1987).

Death-feigning is advantageous to an animal that cannot escape because many predators, including mantids, lizards and cats, strike only at prey which shows signs of life. Some snakes roll themselves into tight, motionless balls, which protects their vulnerable heads when they are attacked. Hognose snakes (*Heterodon nasicus*: Colubridae) sometimes strike themselves sufficiently severely to cause bleeding (Kroll 1977) while the African savanna monitor (*Varanus exanthematicus*) adopts a rigid posture when it feigns death (see Cloudsley-Thompson 1994). Birds also employ thanatosis as a life-saving response and have been known to collapse suddenly in flight and fall to the ground. More usual, however, is injury-feigning, but this is employed, not as a protection for its protagonist but as a means of distracting the attention of the predator from the eggs or young. The stone curlews (*Burhinus* spp.: Burhinidae), for instance, feign injury with remarkable realism when predators invade their nesting areas.

Death-feigning is not peculiar to the American opossums. It is also practised by a number of placental mammals, the most familiar being foxes (*Vulpes* spp.). Ewer (1968) described thanatosis in the ground squirrel (*Xerus erythropus*: Sciuridae), but it is probably not a response commonly employed by mammals. According to Driver and Humphries (1988), injury-feigning, mobbing (Sect. 7.1.2) and other aberrant behaviour may have evolved by a process like ritualization in reverse – one of complexification and apparent disorganization which obscures signals of real value to the reactor. These authors name the development of such random and confusing complexity the "principle of anarchization".

4.3 Deflection of Attack

When attacked by a predator, a prey animal can sometimes escape by inducing its enemy to attack it in the wrong place. Diversion behaviour directs attack away from the vulnerable eggs or young and, in birds, may be achieved by injury-feigning, while deflection marks induce predators to attack non-vulnerable parts of the body where little harm is caused (Edmonds 1974). A classic example of this is seen in the small eyespots on the wings of many species of butterflies. These have long been known to deflect the attacks of birds from the vulnerable bodies of the insects. This hypothesis was confirmed by Carpenter (1941), who noted that the beak marks on the wings of African butterflies were concentrated around the eyespots. Since these spots are normally near the edges of the wings, rather than near the body, the attacks had not been fatal.

Deflection marks can have the disadvantage that they are conspicuous and may actually attract the attention of a predator which had not previously noticed the insect (Edmunds 1974). On balance, however, they must be beneficial or they would not otherwise have evolved. The tails of certain lizards are, likewise, sometimes more colourful than the rest of the body, and are brighter in the young than in the adult. For instance, the tail of *Mabuya quinquetaeniata* (Scincidae) is bright blue in contrast to the black body stripes, but this tail coloration disappears with advancing age. In South Africa, *Eremias namaquensis* (Lacertidae) has a red tail when young: this, too, becomes cryptic later in life (Cott 1940). Lizards often sacrifice their tails in autotomy (Sect. 4.4) and it is better for them that the tail should be attacked by a predator than that their heads should suffer. In *Tiliqua rugosa*, the stumpy tail is very similar in shape and colouring to the head. Automimicry of the head by the tail is also found in some geckoes, snakes, and in amphisbaenians. Not only is the attention of prey distracted from the menace of the true head (Sect. 2.2.1), but the attacks of predators are diverted to a less vulnerable part of the body.

4.4 Autotomy

When attacked by a predator, many insects and arachnids autotomise the limb that has been grasped, and thereby make their escape. The missing leg is then regenerated at the next moult. From experiments on the spider *Kukulcania hibernalis* (Filistatidae), Formanowicz (1990) showed that leg autotomy can be an effective defence against certain kinds of predator. It increases survival in encounters with scorpions but not with centipedes: the differences result from the behaviour and morphology of the two types of predators. Autotomy also occurs in some Chilopoda. When centipedes of the orders Scolopendromorpha and Scutigeromorpha are attacked, the autotomised limbs sometimes show spontaneous movement, slowly bending and straightening to the accompaniment of faint creaking sounds. Only the terminal or anal legs of Scolopendromorpha appear to do this, but the phenomenon appears in all the legs of *Scutigera* spp. (Scutigeridae) which may also stridulate in situ (Dumortier 1964). The anal legs of scolopendrids are not involved in locomotion and show considerable variation in structure. In some cases, this is related to their function in distracting predators, in addition to a possible role as posterior antennae (Lewis 1981). Its seems probable that the attention of the predator is attracted by the movements and stridulation of the autotomised leg, giving the centipede an opportunity to escape (Cloudsley-Thompson 1960a).

Among reptiles, only the tail can be sacrificed in autotomy, which is found in many lizards, some amphisbaenians and a few snakes. The subject of reptilian caudal autotomy has been reviewed in detail by Arnold (1984, 1988) and by Bellairs and Bryant (1985), and was disucssed with particular reference to desert reptiles in an earlier volume of the present Series (see Cloudsley-Thompson 1991). It is probably especially significant for the survival of reptiles in exposed arid environments. At the same time, it is a compromise adaptation involving a sacrifice that cannot easily be afforded, but which is a preferable alternative to almost certain death. Not only do valuable resources have to be expended in regeneration, but individuals that lose their tails suffer decline in social status and, consequently, in reproductive success (Ballinger and Tinkle 1979; Althoff and Thompson 1994).

4.5 Spines, Urticating Hairs and Gin Traps

Both hedgehogs (Erinaceidae) and porcupines (Hystricidae) are successful desert animals not least on account of their spines. When threatened, hedgehogs are able to roll themselves into a tight ball so that their heads, short legs and underparts are protected. Porcupines (Hystricidae) charge backwards at their enemies. Their hind quarters are heavily armed with backward-pointing spines which easily become detached from the body. When lodged in the flesh of a predator, they cannot easily be extracted, and may occasionally even be fatal. The monotreme equivalent in Australia is the spiny ant-eater or echidna (*Tachyglossus aculeata*) (Fig. 17).

Barbed urticating hairs are characteristic of the larvae of Lepidoptera and are found especially in the families Notodontidae, Arctiidae, Lymantriidae, Lasiocampidae and Saturniidae. Single or aggregated secretory cells pour their products onto them so that, when the hairs are embedded in the skin of a potential predator, their toxic contents are released subcutaneously. The urticating hairs of mygalomorph spiders may cause considerable irritation of the skin, but they are not toxic. Arthropods with urticating hairs are not predominant in the desert although some species do occur there.

Fig. 17. Australian echidna (*Tachyglossus aculeata*) burrowing

Lewis (1964) described a form of "gin trap" in *Sternocera castanea* (Buprestidae) – a species of beetle particularly common around Khartoum during the rainy season – which is capable of giving a sharp nip. The posterior border of the prothorax is V-shaped and fits tightly against the closed elytra, which form a corresponding V to receive it. Both the posterior border of the prothorax and the anterior border of the elytra have sharp knife-like edges which accentuate their effect. When the beetle is picked up by the prothorax, a gap of 4–5 mm appears between this and the elytra. After a time lag of 2–3 s, however, the gap is violently closed. The fact that such snapping is only elicited when the body is grasped suggests that the device may be directed against vertebrate predators. The smaller buprestid *Julodis caillaudi* possesses a similar gin-trap. Both pupal and adult gin traps occur in many families of beetles.

4.6 Venoms, Defensive Fluids and Toxins

Venoms are usually regarded as poisonous fluids that are inserted by bites or stings. Defensive fluids are ejected or sprayed at an attacker, while toxins render the bodies of potential prey animals distaseful or poisonous to the predator that consumes them. These categories, however, sometimes overlap, as when an African black-necked or spitting cobra (*Naja nigricollis*: Elapidae) squirts venom at the eyes. Coupled with its large size and the power to eject its venom to a distance of almost 3 m and thus reach the eyes of a standing person, is the fact that this type of attack usually strikes the victim as a complete surprise. Furthermore, it is administered in an instant, the snake rearing and spitting upon the slightest provovation. Again, the urticating hairs of Lepidoptera larvae are venomous but not those of Mygalomorphae (Sect. 4.5).

4.6.1 Venoms

Venomous desert animals include centipedes, Hymenoptera, scorpions and spiders, lizards of the genus *Heloderma* and snakes. The poison claws of Chilopoda insert venom both for the capture of prey and in defence: arthropods and vertebrates are probably the taxa most affected (Lewis 1981). The venom has a haemolytic action resembling that of some spiders and scorpions (Lawrence 1984).

Aculeate Hymenoptera – wasps, bees and ants – are well represented among the fauna of deserts (Crawford 1981). Social wasps use their stings almost exclusively in defence of the colony (Sect. 7.1.1), and unprovoked attacks are comparatively rare (Spradbery 1973). The principal component of the venom of wasps is a protein which produces allergic reactions in people. Vespidae are not the only wasps to use their sting in defence, although it is most effective in them. Solitary wasps which are predaceous and stock their nests with caterpillars, Hemiptera, Orthoptera, spiders and so on – which are stung and paralysed – can also use their stings in self-defence. The same is true of solitary bees although, once again, social species are especially formidable in the defence of their nests. Like the venoms injected by wasps, the stings of bees introduce complex mixtures of proteins and enzymes. The venoms of wasps, however, differ from those of bees in containing a larger proportion of histamine and serotonin (5-hydroxytryptamine).

Ants (Formicidae) possess an enormous variety of offensive and defensive devices which have been summarised by Hölldobler and Wilson (1990). They include mandibles – used for biting and piercing – as well as venomous stings, poisonous droplets and secretions. None of these is unique to desert dwellers, but most of them are employed by species that inhabit arid regions. As with other aculeate hymenopterans, the primitive components of the sting are proteinaceous. They are also neurotoxic and histolytic, and so are crippling to small invertebrate enemies and painful to larger vertebrates. The venom of the fire-ant *Solenopsis richteri* of the southeastern USA has necrotic effects. When a fire-ant colony is disturbed, the ants erupt so quickly that the person responsible may receive as many as 5000 stings in a matter of seconds (Smith 1993).

Scorpion venoms are the complex secretions of apocrine glands. They usually contain multiple proteins with low molecular weights – the neurotoxins – mucus, salts, oligopeptides, nucleotides, amino acids and other organic compounds. Over 70 different toxins have so far been isolated from the venoms of various species of scorpions, although the presence of multiple toxins in a single venom has not been satisfactorily explained (Simard and Watt 1990). Because the active ingredients specifically target neuronal membranes, which they short-circuit, they cause depletion of the chemical transmitters used to communicate between nerves and muscles. The result is convulsions and paralysis in both invertebrates and vertebrates (whose nervous systems are not markedly different). Minor alterations of an invertebrate toxin can convert it to a vertebrated toxin. The scorpion sting thus has a dual purpose – it is an offensive weapon against invertebrates and is defensive against vertebrates.

The world's most venomous desert scorpions are *Androctonus* spp. of North Africa and the Middle East, which can deliver massive doses of neurotoxins, and *Centruroides* spp. in North America. All these are Buthidae, as is *Leiurus quinquestriatus*, which produces the most toxic venom known, although in smaller quantities. These have relatively small and slender claws. Scorpions with massive claws (Chactidae, Scorpionidae, Diplocentridae and Bothriuridae) are nearly all relatively harmless while members of the more dangerous families (Buthidae and Vaejovidae) have slender chelae (Cloudsley-Thompson 1984b, 1993b).

Spider venoms are not primarily defensive. Few spiders have chelicerae sufficiently powerful to penetrate mammalian skin and in even fewer is the poison effective against vertebrate predators. A notable exception is provided by the genus *Latrodectus* (Theridiidae) which includes the cosmopolitan black widows (*L. mactans* and *L. geometricus*), the malmignatte (*L. tredecimguttata*) of the Mediterranean region, and the "shoe button" spider – as *L. mactans* is called in South Africa. In Australia it is known as the redback. These spiders all have an extremely toxic venom, but they do not readily bite in self-defence. The North American brown recluse (*Loxosceles rufescens*: Sicariidae) is another dangerously venomous species of arid lands. Its poison has a necrotic effect similar to that of fire-ants. It is widely distributed, and has been introduced into Australia. Most other species of spiders that can be harmful to vertebrate animals are found in the more humid regions of the world (Cloudsley-Thompson 1993b).

In contrast to that of snakes, the venom apparatus of the aposematic Gila monster (Fig. 18) and Mexican beaded lizard (Sect. 2.1) consists of four pairs of deeply grooved and flanged teeth in the lower jaw, with

Fig. 18. Gila mouster (*Heloderma suspectum*). (Cloudsley-Thompson 1994)

ducts from the venom glands opening near their bases. Ponderous and sluggish, these slow-moving reptiles prowl at dusk, feeding on nestling birds and mammals, eggs and lizards, so their venom probably has a defensive rather than a predatory function. Venomous snakes of arid regions are back-fanged Colubridae, most of which chew venom into the wound caused by their bite, Elapidae, of which cobras are the most important, and Viperidae. The last-named family includes the subfamilies Viperinae – the true vipers – and Crotalinae – the pit-vipers and rattlesnakes.

In most cases, a snake strikes rapidly in the darkness and then disappears before it can be identified. The characteristic punctures made by the fangs, however, often enable diagnosis to be made: the presence of two bleeding fang marks is a clear indication that the bite is that of a dangerously poisonous species. If other tooth marks are also present, a cobra is responsible; if not, the bite will have been inflicted by a viper. Cobra venom acts rapidly on humans and, if death from respiratory failure does not occur within about 12 h, the patient normally recovers quickly. In the case of viper bites, however, death is less rapid, and it may be several days before the patient is out of danger because late complications, such as haemorrhage and sepsis, are not infrequent.

Although the venom glands of snakes have been derived from salivary glands, it is not necessary for the prey to have been poisoned for it to be digested. Snakes force-fed with rats or birds can digest them quite adequately. Venom may assist active digestion, but it is by no means essential in the case of snakes, as it is to spiders and centipedes. Indeed, in some snakes it seems that poison has become far more important in defence than it is in offence. Despite their raised fur, which may afford some protection, and partial if not complete immunity to venom, mongooses depend largely upon their speed when attacking cobras and vipers (Hinton and Dunn 1967). Anyone who has witnessed a fight between an Indian mongoose (*Herpestes edwardsi*) and a snake will have been impressed by the intense concentration of the little mongoose from the moment it sees its adversary, its aggressive pugnacity, and the incredible rapidity of its movements (Fig. 19).

Fig. 19. Indian mongoose (*Herpestes edwardsi*) attacking a snake

4.6.2 Defensive Fluids

Woodlice (Isopoda: Oniscidae) are endowed with five or six types of tegumental glands of which the lobed glands, located on the lateral plates and uropods, open to the exterior through minute pores (Warburg 1993). The degree of development of the lobed glands varies greatly among different species. They secrete a strongly acidic secretion with a repellent odour, which is believed to be a defence mechanism against scorpions (Herold 1913) and spiders (Gorvett 1956).

In addition to the use of the poison claws in defence, centipedes produce a number of defensive secretions. These have been discussed in detail by Lewis (1981). Millipedes, likewise, secrete defensive secretions which are certainly repugnatorial and may even cause blindness if they get into vertebrate eyes. Millipedes usually discharge their secretion only from segments that are threatened and are thus protected from repeated attacks by small predators such as ants (Hopkin and Read 1992). Usually, they simply ooze out onto the body surface, but some large tropical species are able to eject their defensive fluids for considerable distances when attacked by vertebrates; but this does not prevent them from being devoured by certain enemies such as meerkats and banded mongooses (*Mungos mungo*; Lawrence 1984).

Defensive secretions are produced by many insects and arachnids. The major invertebrate enemies of termites appear to be predatory ants. In some termite genera, such as *Nasutitermes* and *Trinervitermes*

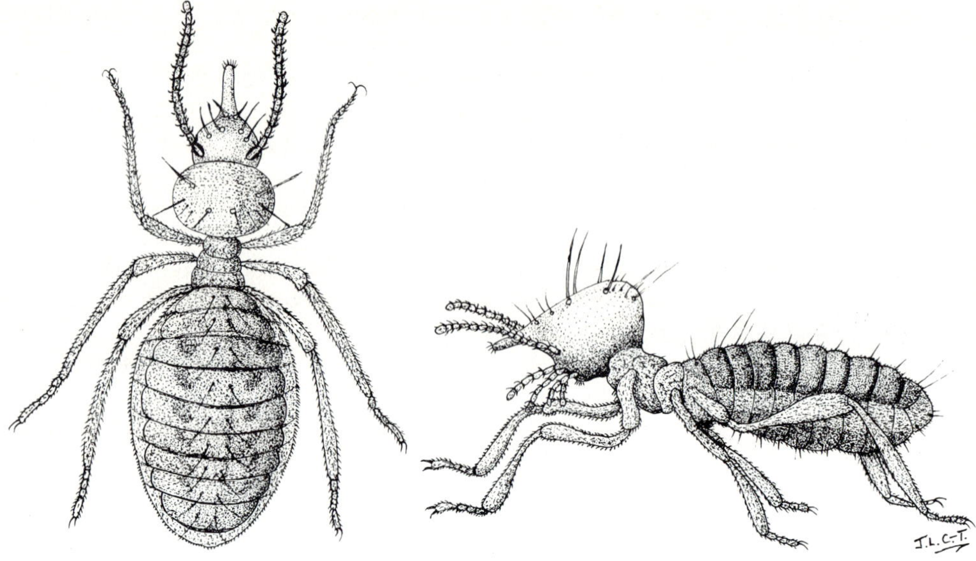

Fig. 20. Nasute termite soldier (Nasutitermitinae)

(Termitidae), the soldier caste is adapted for chemical defence. The mandibles are artrophied and the head is drawn out into a tubular "nose" terminating in the opening of the frontal gland, from which a sticky fluid is ejected upon any attacker (Fig. 20). The defences of Sonoran desert termites have been reviewed by Jones and Nutting (1989). They include both chemical and mechanical defences: biting, regurgitation, defaecation, abdominal dehissence, walling-off intruders and the formation of elaborate structures. Other insects that excrete glandular exudates or allomones include Collembola, Thysanura, Dermaptera, Dictyoptera, Hemiptera, Neuroptera, Lepidoptera, Diptera, Hymenoptera and Coleoptera (reviewed by Whiteman et al. 1990). Certain caterpillars and beetles that do not use defensive sprays air their defensive fluids by evaginating the entire gland that secretes them (Eisner and Meinwald 1966).

Beetles are prominent among chemically defended arthropods, especially Chrysomelidae, Staphylinidae and Carabidae. The latter include the bombardier beetles (*Brachinus* spp.), which discharge a hot toxic mixture at attacking predators. Most desert Tenebrionidae are smelly and presumably distasteful, while *Eleodes* spp. spray quinones in defence (Sect. 2.1). Many insects contain toxic or distasteful allomones within their bodies. Some bleed reflexively when attacked; certain moths eject a mixture of blood and cervical gland secretions, and grasshoppers discharge their gut contents when molested. Other insects defaecate, while some reduviid bugs spray enzyme-rich salivary secretions with accuracy up to 30 cm. Serious lesions may result when these contact the sensitive eyes of vertebrates (Edwards 1961).

Arachnids that produce defensive compounds de novo [in contrast to the sequestration of plant metabolites (Sect. 7.3)] include Opiliones, which secrete a great variety of ketones and quinones from paired cephalothoracic defensive glands, Thelyphonida, which spray an acid secretion from glands opening near the rear end of the body (Eisner 1962; Eisner and Meinwald 1966), and spiders of several mygalomorph families, which turn their rear ends towards the enemy and squirt from the anus a clear liquid that may have irritating properties (Main 1976; Cloudsley-Thompson 1995). The black widow spider (*Latrodectus hespherus*) emits an adhesive, viscous web from its spinnerets when threatened by predatory enemies such as mice (*Peromyscus* spp.) and, if necessary, places this on the offenders. The deterring effect of the silk is due solely to mechanical irritation (Velter 1980).

Defensive glands may be used singly or in groups. They are sometimes extremely complex. For example, *Piezodorus teretipes* (Pentatomoidea), a green stink bug, very plentiful in the Sudan at the time of the annual rains between July and September, secretes *trans*-hex-2-enal, which both smells and tastes unpleasantly peppery and oily (Gilchrist et al. 1966). Arthropod secretions often include more than one type of quinone accompanied by other substances. The alarm pheromones of ants may also have a repellent function: there is a parallel here with plants which secrete the same insect-repelling substances, such as terpenes, that insects themselves use in an anti-predatory role. The chemical defences of arthropods have been reviewed by Blum (1981), who stated that more than 110 different hydrocarbons are produced in their various exocrine glands.

Desert vertebrates that secrete defensive fluids include toads (Duellman and Trueb 1986), reptiles (Guibé 1970) and mammals (Edmunds 1974). Many of these, however, do not have specific repugnatorial glands, but the entire flesh or body fluid is noxious. A classic example is afforded by birds, the edibility of whose flesh is correlated with conspicuousness. Distasteful species are often aposematic, whereas edible species tend to be cryptic. In October, 1941 Cott (1947) noticed that hornets were attracted to the flesh of a palm dove (*Streptopelia senegalensis*: Columbidae) which he had skinned, whilst ignoring that of a pied kingfisher (*Ceryle rudis*: Alcedinidae). In further tests, carried out in Egypt and Lebanon, both hornets and the domestic cat played the part of taster, and it was confirmed that the majority of conspicuous species were distasteful, the exceptions possessing other means of defence or little subject to attack.

The results of natural selection are perhaps nowhere better demonstrated among animals than in the evolution of defensive substances and the means by which they are deployed.

4.7 Warning Smells, Sounds and Deimatic Display

It is naturally beneficial to any animal to avoid attack, however well it may be armed. Consequently, venomous animals frequently have aposematic colours which serve to warn potential enemies to avoid them (Sect. 3.8) or, if primarily cryptic (Sect. 3.4), may display warning colours when their first line of defence, crypsis, has been breached and they are threatened. For example, desert vipers tend to be cryptic, thereby avoiding the attention fo hawks, secretary birds (*Sagittarius serpentarius*: Sagitariide) and their ecological analogues, road runners (*Geococcyx californianus*: Cuculidae), mongooses and so on. When disturbed, however, many of them display warning colours or scents, or emit threatening hisses. The venomous Gila monster (Sect. 4.6.1, Fig. 19) does the same. Normally inconspicuous in the dappled shade beneath bushes, it has striking black and yellow aposematic coloration when viewed in the open.

Guilford (1992) proposed that distinctive odours of aposematic prey might enhance the ability of a predator to associate conspicuous coloration with unpalatability. This has been demonstrated experimentally to be the case with the plains garter snake (*Thamnophis radix*: Colubridae), but with a twist. In garter snakes, conspicuous coloration strengthens the learning of chemosensory characters, but not vice versa. The chemosensory potentiation of visual learning, as proposed by Guilford, may be characteristic of birds and other visually oriented predators, while visual facilitation of chemosensory learning is more likely to be important in the case of predators, such as snakes, which rely more heavily upon olfactory cues (Terrick et al. 1995).

Threatening or deimatic behaviour (as in *Brachysaura minor*: Agamidae; Fig. 21) may give genuine warning that an animal can attack and harm the predator, or it may be pure bluff, but there is no clear distinction between the two. To one enemy the display may be a genuine warning, but to another it may just be bluff (Edmunds 1974). The Namib desert chameleon (*Chamaeleo namaquensis*) is unusual in that it can run surprisingly fast, hides in rodent burrows and also digs burrows for itself. If chanced upon whilst at rest, it keeps perfectly still – except for the eyes which rivet themselves upon the enemy. When threatened or attacked, however, the chameleon leaps to a high, stiff-legged stance, quickly becomes black, inflates and compresses its body laterally. The throat becomes gorged, showing the yellow, reddish-orange interstitial skin, and the mouth is held widely agape while the animal emits hissing sounds and gutteral growls. The mouth of the adult has a yellow-orange interior, but that of the young is black. In its display, the desert chameleon turns towards and rushes at its enemy. Such action is usually quite startling. In moments of hesitation on the part of the antagonist, it may

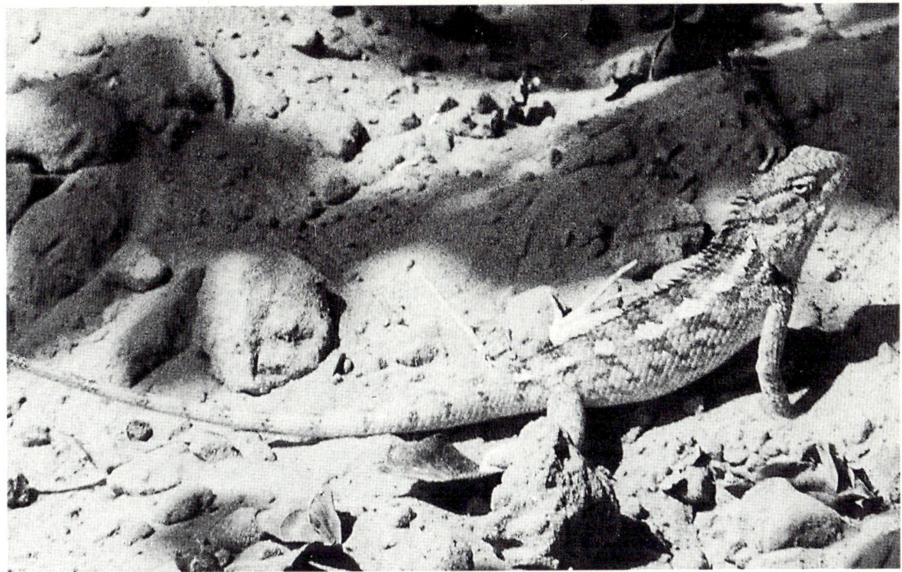

Fig. 21. Deimatic display of an agamid lizard (*Brachysaura minor*) (Thar desert)

change to a lighter colour and attempt to escape. It is most vicious if thwarted, however, making repeated charges and snapping its powerful jaws. Not only can it inflict a painful bite, but it does not release its hold: instead, it twists and turns, pulls and pushes, as it drives its teeth into the flesh of its adversary (Burrage 1973). Deimatic behaviour has been discussed by G. Costa (1995), who points out that some false eyes, suddenly demonstrated, may not only serve the deflect attack (Sect. 4.3) but may also have a deimatic function.

Warning smells, given off only when under threat, include those of various insects. They are generally offensive and have an advertising function comparable to that of aposematic coloration. Some venomous snakes, such as cobras, produce chemical warning signals which prevent them from being attacked by potential enemies such as monitor lizards (*Varanus* spp.). They differ from defensive fluids (Sect. 4.6) according to the classification adopted in the present volume, because they are not themselves necessarily the defensive mechanism to which they are alerting the potential aggressor, but may be associated with physical or chemical weapons (Edmunds 1974). Porcupines and hedgehogs, for instance, although defended by their sharp spines (Sect. 4.5), can also produce unpleasant smells (Cott 1940). In most cases, however, an evil-smelling stench is a defence in itself, and is enough to deter most aggressors. Garter snakes (*Thamnophis* spp.) discharge foetid, milky secretions from the glands of the cloaca. These unpleasant substances cannot be projected at an enemy, like the chemicals produced by many insects and mammals, but they quickly smear the body of any animal that presses its attack.

Warning sounds include the stridulations of mutillid wasps or velvet ants (Mutillidae), scorpions (A. J. Alexander 1958, 1960; Constantinou and Cloudsley-Thompson 1984; Lourenço and Cloudsley-Thompson 1995), Solifugae (Cloudsley-Thompson and Constantinou 1984), saw-scaled vipers (*Echis* spp.), horned vipers (*Cerastes cerastes*) and the sand viper (*C. vipera*) of North Africa (Le Berre 1989). In these, the orientation of the lateral scales of the trunk has changed so that their keratinous keels are inclined with the posterior edges downwards. The snakes can therefore stridulate without losing respiratory water, as they would do in hissing, by rubbing these scales against each other (Mendelssohn 1963). *Scorpio maurus* emits a sound by striking the ground with the distal portion of its metasoma (Laurenço and Cloudsley-Thompson 1995). The acoustic characteristics of the rattling sounds of rattlesnakes (*Crotalus* spp.) have been analysed by Fenton and Licht (1990), who found that they all conform to the same general pattern. They are medium-intensity, broad-based sounds with rapid onset and no structured change in frequency pattern over time. Therefore, they must be a signal designed to attract the attention of other animals, whose responses will reflect their hearing characteristics and previous experience. While the majority of snakes are silent, there are a few which hiss in characteristic ways and others that vibrate their tails against vegetation. Sweet (1985) suggested that the tail vibrations of non-venomous gopher snakes (*Pituophis melanoleucus*) producing an audible buzz in dry leaves, may be mimicking the sounds of rattlesnakes (*Crotalus viridis*). Deimatic displays are not common in birds, since they tend to avoid their enemies by flight, but the nestlings of burrowing owls (*Speotyto cunicularia*) also make sounds that resemble the rattles of *Crotalus* spp. (Maclean 1996). In several species of *Pituophis* there is a development of the thorax which produces a particularly loud hiss.

Porcupines are strictly nocturnal, but the whiteness of their quills, when erected in a characteristic fan, renders the animals conspicuous in the dark. Furthermore, when hard-pressed by an enemy, the porcupine shakes its tail quills, which are specially modified to form a rattle, utters hoarse and grutteral sounds, and produces a powerful stink. Only if none of these warning devices in sufficient to secure the departure of the enemy will the porcupine launch itself backwards and deal the aggressor a blow with the barbed quills of its tail (Sect. 4.5; Cloudsley-Thompson 1980).

An apparent increase in body size is a deimatic device employed by amphibians, reptiles such as chameleons, lizards, and some snakes, birds and mammals. Frogs and toads inflate their bodies with air, agamid lizards reach upwards with extended fore-legs (Fig. 19), exposing large teeth which can inflict exceptionally painful bites if necessary (Sect. 4.8).

Deimatic display among desert animals is reviewed in greater detail in another volume in this Series (G. Costa 1995).

4.8 Retaliation

When flight is impossible, the only possible alternative may be to fight back. Defensive response and degree of aggressiveness is inversely related to the speed at which a lizard can sprint. At high temperatures, both *Agama pallida* and *A. savignyi* (Agamidae) flee rapidly from predators. At lower body temperatures, however, they seldom run, but, instead, hold their ground and attack aggressively. The shift from flight to fight is almost certainly adaptive, because a cold lizard would have little chance of out-running and out-manoeuvring natural predators in the Middle East and Negev deserts where these species occur (Hertz et al. 1982). When the Arabian scaly-tailed agamid *Uromastyx microlepis* is cold, it has a conspicuous dark slate-grey colour but, as it warms up, it becomes deep yellow and then matches its sandy background. Cold lizards do not flee, but puff themselves up (Sect. 4.7), exhale with loud, threatening hisses, and swing their large, spiny tails, which are most effective weapons (Cloudley-Thompson 1991, 1994). Indeed, second to teeth, tails are among the most effective weapons of defence among reptiles.

Most animals attempt to escape rather than fight. Scorpions move slowly, backing towards their refuges and, if the aggressor approaches, they lash at it with their tails. Insects bite or sting with varying degrees of readiness, but Solifugae and spiders show great reluctance to use their chelicerae in defence. This is true of spiders, such as *Latrodectus* spp., which produce venoms that are extremely toxic to vertebrates. Small birds often retaliate by mobbing predatory raptors, and even apparently harmless mammals, such as rodents, will bite if cornered, while antelopes use their horns and hooves. Indeed, antelopes and other ungulates with light, desert-coloured coats have dark horns which draw attention to their weapons of defence and give warning to potential aggressors not to attack (G. Costa 1995).

Horns, like antlers, probably evolved mainly as weapons of intraspecific aggression. They are found in Giraffidae, Cervidae and Bovidae. The last-named family includes the North American pronghorn (*Antilocapra americana*) as well as numerous species of antelopes, gazelles, wild sheep and goats and other Artiodactyla of arid regions. This primary function of horns is to prevent the heads of rival males from slipping during intraspecific territorial combat (Sect. 7.1.4). As defensive weapons they are often effective, and cases are known in which even lions have been killed by gemsbok and buffalo (*Syncerus caffer*)

(Bubenik 1990). It is surprising, however, how little defence is put up by many antelope once they have been caught. Lions kill smaller prey with a quick blow of the paw, finishing them off with a bite in the neck or throat. Animals the size of a wildebeest are usually thrown down by the impact of the lion hurling itself against them. With one paw, the attacker often catches hold of the forehead or the nose of its victim, pulling the head down to the chest so that, in falling forward, the animal breaks its neck (Schaller 1972).

Lions usually focus their attack on the front part of the body of the prey. Exceptions occur, however, when lions combine to attack a large animal, such as a buffalo or a giraffe. On these occasions, whilst one or more lions attack from the front, another will grab the hind legs of the prey from the rear. Lions double their chances of success by hunting cooperatively, hyaenas quadruple them (Schaller 1972). In the Kalahari, where gemsbok with their dangerously sharp horns form the staple prey of lions, kills are invariably made by leaping upon the victim and breaking its back. The strength of a full-grown lion in its prime is prodigious. He can fell an ox with a blow of his paw, crush the bones of its neck in a moment, and drag the whole carcass to his lair. By exerting his strength, a lion can even drag an animal the size of a buffalo for 50 m over flat ground but, of course, he cannot lift it much above the surface.

Not surprisingly, the reaction of most artiodactyls and giraffes to the attacks of lions, leopards or hyaenas is to flee. Only when this becomes impossible, or when defending their young, will they fight with horns and hooves. Indeed, the kick of a giraffe can easily maim or kill a lion: the horns are used only in sexual and territorial combat. G. Costa (1995) discusses retaliatory behaviour among desert animals in further detail. Theories of predator-prey oscillations have been analysed by many authors (see Pianka 1978) and will not be discussed here. It is evident, however, that in arid open country many of the secondary anti-predator devices are found that have been exploited by animals in other biomes.

5 Parasitic and Allied Interactions

No more intimate relationship exists between one species and another than the bond linking parasites with their hosts. Parasites have traditionally been regarded as having an impact on the populations of their hosts rather similar to those of predators on prey populations. Recent studies, however, indicate that the influence of parasites is more subtle than that of predators. They may kill their hosts or have only slight, almost undetectable effects on their viability. They can alter the fecundity of the host, its competitive ability, physiology, social dominance or its ability to attract mates. They may even be beneficial by making intermediate hosts more available as prey, or by reducing the fitness of co-infested competing species (Bull and Burzacott 1993).

Insofar as parasites themselves are concerned, the main problem lies in achieving transmission from one host to another. This difficulty is especially pronounced among desert species because individuals tend to be widely dispersed and are often inactive, or in a state of cryptobiosis, diapause, or dormancy, for long periods of the year. The life cycles of parasites must, for that reason, be closely synchronized with those of their hosts. This is a major aspect of the lives of the parasitic plants and animals in arid lands, and one with which the present chapter is concerned.

In some cases, however, transmission presents no greater problems to the parasites of animals in arid regions than it does elsewhere. Indeed, if may even be facilitated by the localised distribution of microclimates favoured by the hosts. As everyone who has travelled in the desert will know, the ground beneath trees and other forms of shelter, especially in wadi beds, is frequently heavily infested by larval ticks, awaiting the arrival of some unfortunate mammalian host (Sect. 5.2). In the case of amphibians that aestivate or hibernate for long periods, the transmission of monogenetic trematodes also needs to be synchronised with the activity of their hosts (Sect. 5.3) when the latter congregate whilst reproducing in temporary rainpools.

5.1 The Parasites of Plants

As far as plants are concerned, conditions in the desert are inimical to life, not only of the vegetation but also of the pests and diseases that affect it. Indeed, the parasitism of *Cistanche phetypara* (= *tinctoria*)(Orobanchaceae) upon the roots of *Tamarix* spp. may possibly, in the first instance, serve the purpose of securing water, since salts absorbed by the roots of the latter are, in part, excreted by salt glands on the reduced leaf surfaces of the tamerisk (Grenot 1974). *C. phetypara* is a large, robust, perennial, root parasite common not only on tamarisk. In the littoral salt marshes of the Gulf region it is also found on *Limonium axillara* (Plumbaginaceae) and *Arthrocnemum glaucum* (Chenopodiaceae) (Batanouny 1981). Although parasites and pathogens often exert a great impact on populations, both of plants and of animals, their influence in desert communities is little understood.

Conditions in the surrounding desert often act as a barrier to the introduction into oases of crop pests and pathogens. When they do become established, however, usually as a result of human activity, both pests and diseases tend to flourish. For example, the scale insect *Parlatoria blanchardi* (Coccoidea), an important pest of date palms, was probably introduced into the Sahara long before the Arab invasion. It is now widely distributed throughout North Africa, but has not yet reached Chad and the western oases. It was, however, accidentally introduced into Colomb-Béchar in 1920, certain oases of Touat in 1912, and Tata in Morocco as recently as 1945. This species and another, *Phoenicococcus marlatti*, were inadvertently introduced, along with their host plant, into Arizona and California: it took over 30 years to eradicate them (see Cloudsley-Thompson 1974a).

The majority of mites that feed on the tissues of higher plants belong to the order Prostigmata. The Tetranychoidea and Ercophyidae are exclusively phytophagous while the family Tarsonemidae, although including a number of mycetophagous forms, also contains several important phytophagous species. Many plant-feeding mites live in colonies of various sizes on the stems and foliage, or in the buds of the host plants, and associated with these aggregations are a number of acarine predators, particularly members of the family Phytoseiidae (Mesostigmata; Evans et al. 1961).

5.2 Ectoparasites

Although they are restricted to the host's first line of defence, the integument, away from the internal organs, ectoparasites may have a pronounced effect on the fitness of the host. They may cause loss of weight, milk or eggs in the case of mammals and birds respectively, foetal abortions and even death, in addition to transmitting pathogens. Different species may share the exploitation of their host's integument and their great diversity reflects the independent origin of the numerous taxa involved – for instance at least seven orders in insects alone. In order to assess the impact of parasites on the fitness of their hosts, it is necessary to manipulate the experimental load experimentally; but negative correlations between parasite load and components of host fitness do not prove that parasites have no impact. Ectoparasites may have hidden impacts associated with the defences of the host. For instance, birds may desert their nests even though eggs or young are present. Indeed, the fact that, unlike endoparasites, ectoparasites are often capable of escaping from dead hosts means that selection does not favour the establishment of tolerance between parasite and host to the extent that it does in the case of endoparasites (Lehmann 1993).

There are many ectoparasitic taxa in the desert, and they can make obvious contributions to the structure of food webs (Crawford 1981, 1991a, b). The presence of rodents in sandy deserts has influenced their ectoparasites. For instance, in Asian dunefields, burrows harbour special communities including fleas, which have evolved closely with their hosts (Krivokhatskiy 1985). Parasitic insects are found wherever their hosts occur, and Polis (1991b) has documented their involvement in the Coachella valley foodweb.

Desert arthropods and vertebrates of all taxonomic groups are frequently infested with mites. Dung beetles (*Copris* spp.: Scarabeidae) almost invariably carry numerous parasitic or symbiotic mites on their bodies. For instance, *C. hispanus* is normally infested by *Parasitus copridis* (Mesostigmata). The advantages of this close association are clear as fas as the mites are concerned: *Copris* beetles are attracted to fresh dung and, by their activities, they depress the formation of mould which seems to be harmful to mites. A rich fauna of nematodes, collembolans and other arthropods forms at the interface between the dung and the soil. These serve as food for the mites. Futhermore, the mites are always carried by the beetles to an optimal habitat, away from heat and drought, in brood pills and dung heaps (M. Costa 1964).

In a similar way, large infestations of mites are found on burying beetles (*Necrophorus* spp.: Silphidae). The main food of these mites (in temperate regions) appears to be the eggs of Diptera. When an adult *Necrophorus* arrives at a carcass, the mites on it disperse and destroy any fly eggs present, thereby usurping the cadaver for the nutrition of the beetles' own young. If, however, calliphorine larvae are already present in large numbers, *Necrophorus* may not be able to breed (Springett 1968). The same situation may well also apply in desert regions where burying beetles are common. Thus, whereas infestation of *Copris* spp. by *P. copridis* is clearly an example of phoresy, as only the mites benefit, the association between *Necrophorus* spp. and mites should perhaps be classed as symbiosis because both partners benefit from the presence of the other.

Like other arthropods, scorpions are host to mites as well as to nematodes, but there are no reports of them being the hosts of parasitoids, as spiders and Solifugae may be (Sect. 5.4). The pectines and membranes at the articulations of the exoskeleton of scorpions are particularly susceptible to parasitism by mites, whose primary feeding sites vary in different families (McCormick and Polis 1990). *Pimeliaphilus isometri* (Pterygosomidae) parasitises scorpions while *P. podapolipophagus* infests cockroaches, but these are exceptions: all other known species of the genus are parasitic on lizards.

The biological relationships between mites and other invertebrates show varying degrees of complexity. Trägårdh (1943) recognised five major categories as follows: (1) parasitic, (2) predatory, (3) exudate feeders, (4) phoretic species and (5) commensals with social insects. All of these occur in the desert fauna. Parasitic and predatory mites form well-defines groups but, in the case of species participating in apparently harmless associations, data are lacking on the true relationships. Some ectoparasites, such as *Dinothrombium* spp. (Trombidiidae) are parasitic in the larval stages only. These attach themselves to locusts and grasshoppers: the adults prey on termites (Tevis and Newell 1962). Endoparasitic mites are less common than extoparasites, but are not infrequent in the tracheal system of bees. In parasitic mites, the chelicerae may be hooked or needle-like. Hooked chelicerae are found in the parasitic larvae of the Trombiculiidae and Erythraeidae and form a feeding tube; needle-like chelicerae are well adapted for piercing the skin of the host, and occur in both ectoparasites and endoparatistes (Evans et al. 1961).

The effects of ectoparasitic infections may be direct, harming the viability of the host, or they can be indirect when harmful endoparasites are transmitted (Lehmann 1993). Between 1982 and 1990, Bull and Burzacott (1993) conducted a study of the effects of infestation by ticks (*Aponomma hydrosauri* and *A. limatum*: Ixodidae) on natural populations of stumpy-tailed lizards (*Tiliqua rugosa*) in South Australia. They found that lizards with high tick loads in one year tended to have high loads in the next. Longevity was either not correlated with tick load, or was positively correlated.

This was probably because individuals successful in established home ranges regularly revisit well-protected refuges in which tick populations are high. A similar explantation could well apply to the positive correlations between tick load and the size achieved, and between tick load and reproductive success (estimated indirectly by whether or not the lizards were found in pairs in spring). The authors concluded that their data do not support the hypothesis that, in this instance, tick load diminishes host fitness.

In contrast, Sorci et al. (1994) found that, in common lizards (*Lacerta vivipara*), whose range includes much of Europe and northern

Asia, a heavy load of parasitic mites (Laelapidae) increases the speed of sprinting of females. Furthermore, the male offspring of parasite-free female lizards are more likely to survive and disperse than are those produced by parasitized mothers, whereas the opposite is true of female offspring. The results thus support the hypothesis of local adaptations: organisms living in variable environments may suffer dramatic losses of fitness in consequence of fluctuations in the quality of the habitat. These can be combatted either by the organism being phenotypically plastic, or by its searching for a more favourable environment. The adoption of either one of these strategies depends upong the associated costs and benefits, and is based on the assumption that mites and ticks have a direct negative impact on host fitness (Lehmann 1993).

Natural processes invariably interact in more ways than one. Not only may reptiles be camouflaged, but so also may their ectoparasites. For instance, the lizard tick (*Aponomma exornatum*: Ixodidae) of eastern Africa is often found in considerable numbers on the Nile monitor (*Varanus niloticus*). The ticks attach themselves to various parts of the body, especially the axilla, the groin or above the vent. They are handsomely ornamented, with a pattern of yellowish spots on a blackish brown background which harmonizes closely with the coloration of their host (Fig. 22). Snake ticks, likewise, such as those of the black and white cobra (*Naja melanoleuca*), may be well camouflaged. Darker-coloured ticks can usually be seen on the black scales, pale ticks on the white

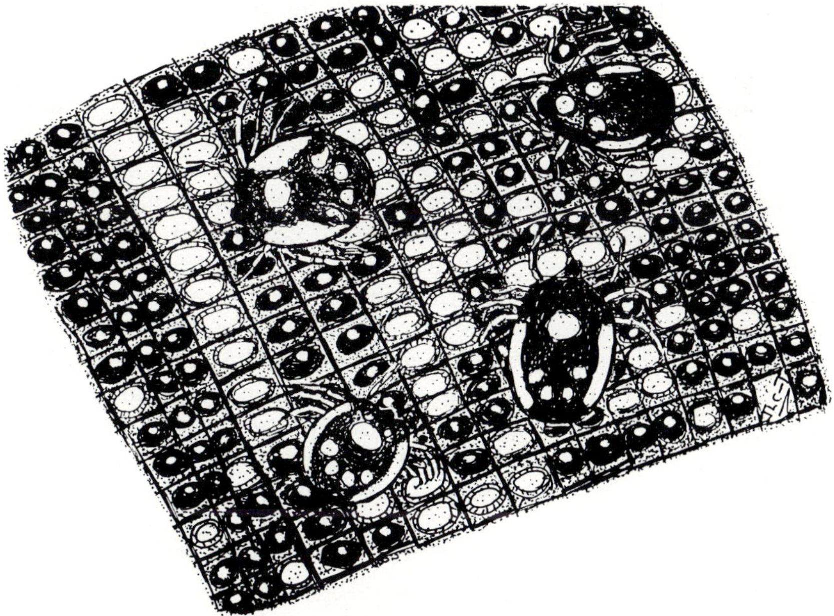

Fig. 22. Lizard ticks (*Aponomma exornatum*) on the scales of a monitor lizard (*Varanus niloticus*). (Cloudsley-Thompson 1994)

scales of the snake. This presumably indicates that the ticks themselves are endowed with colour vision - unless individuals that attach themselves to scales that do not match their own colours get removed. It seems unlikely that the colour of the scales should have an effect upon the colour of the attached tick. Reptiles are not known either to groom one another or to remove their own ectoparasites. The advantage to ticks of matching the colours of their hosts could, however, lie in the fact that reptiles with conspicuous parasites might become particularly vulnerable to predation (Cloudsley-Thompson 1994).

As in the case of *Aponomma exornatum*, mentioned above, *A. hydrosauri* is found mainly under the fore limbs and on the middle of the back of *Tiliqua rugosa*, whilst *Amblyomma limbatum* attaches itself to the regions of the ears and lower back. Data from field surveys suggest that higher proportions of female *Aponomma* occur in both preferred sites and of *Amblyomma* on the head, neck and ears (Chilton et al. 1992).

Many lizards are endowed with mite pockets. These are small invaginations of the skin in places, such as the neck, axilla, groin and post-femoral region. They frequently contain feeding chiggers - the larvae of trombiculid mites - which, in general, are much more common on species with pockets than on species without. Mite pockets have evolved many times and are found in at least five families - Iguanidae, Chamaeleonidae, Geckkonidae, Lacertidae and Scincidae. Arnold (1986), who studied them in detail, believed that they evolved in forms prone to infestation by trombiculids, and serve to ameliorate the damage that this causes by concentrating the chiggers in places that are equipped to minimize the harm that they do.

Now, Trombiculidae sometimes occur in patches which are quite noticeable, especially when the mites are bright red. Giant velvet desert mites (*Dinothrombium* spp.) are scarlet and distasteful to scorpions, Solifugae and other predators (Cloudsley-Thompson 1962), but there is no direct evidence that chiggers impart distastefulness to their hosts. Moreover, mite pockets make the parasites less easily seen, for they tend to be situated in sheltered parts of the host's body and enclose the mite larvae. It can only be concluded, therefore, that the red coloration of chiggers is an atavistic legacy, and that it is of no selective importance, either to the mites themselves or to their hosts. Unlike ticks, trombiculid mite larvae are extremely small, so their coloration is unlikely to be ecologically significant, although it might perhaps have some physiological function (Cloudsley-Thompson 1994).

Parasitism of the barn swallow (*Hirundo rustica*) by blood-sucking mites (*Ornithonyssus bursa*) also has an effect on reproduction. It has been shown experimentally that incubation periods are longer, and nesting periods shorter, if nests contain large numbers of mites. When the parasites were removed by spraying with insecticide, the intervals between clutches were significantly shorter because the nests were re-used for second clutches and the adults were less heavily infested. (Fewer second clutches were laid in nests containing many mites). Moreover, mate choice affects the level of parasitism. Mites, therefore, cause loss of fitness to their hosts, although swallows are not without counter measures and can move to other nests. Thus, there is a "trade-off" between the costs of infestation and of moving (Møller 1990). In contrast, Lee and Clayton (1995) concluded that bird lice (Mallophaga) and louse flies (Hippoboscidae) do not appear to affect reproductive success and survival of the swift (*Apus apus*: Apodidae).

Flocking birds naturally suffer more from parasites in general than do territorial species, and disease is more likely to break out when population densities are high and there is competition for food (Ford 1989). Acarine communities in the nests of birds and small mammals include ectoparasites and other epizoic forms as well as free-living scavengers and their predators. Many of them have an extremely unspecialized diet, and can feed on blood, faeces, mite eggs and so on with equal alacrity, but the majority are scavengers that eat fungi, detritus, excrement and the remains of dead animals (Evans et al. 1961).

Ectoparasitism may take various forms: the parasites may live on the surface of the host, feeding upon exudations and abrading the skin, or they may invade the surface tissues, attack living cells and drink tissue fluids. Fleas, lice, ticks and other mites – no terrestrial vertebrate is able to escape their attentions (Fig. 23). The ectoparasites of camels include mites and ticks. Mange is caused by the mite *Sarcoptes scabiei cameli* (Sarcoptidae) and is very contagious, while ticks, mainly of the genera *Rhipicephalus*, *Amblyomma* and *Hyalomma* (Ixodidae), are important ectoparasites that cause considerable irritation to their hosts. Larvae of the camel nasal fly (*Cephalopina titillator*) are important parasites of the nasal and frontal sinuses of camels. They occur almost universally in Ethiopia, Sudan, and Arabia, and probably everywhere the camel is found (R. T. Wilson 1984). Chacma baboons (*Papio ursinus*) in the Kuiseb River canyon, Namib desert (Brain 1990; Sect. 2.1) are always infested by ticks (*Rhipicephalus* spp.; Fig. 24) which regularly attach themselves to the animals' ears. Each time these ticks are pulled off, a piece of flesh is also removed attached to the hypostome. Eventually the baboons' pinnae almost disappear! Very little is known of the importance of ectoparasites to wildlife in the desert, or to what extent they affect population numbers. Susceptibility of impalas (*Aepyceros melampus*) to ticks and lice (Anoplura) in semi-arid regions does, however, appear to increase during periods of drought (Horak et al. 1995).

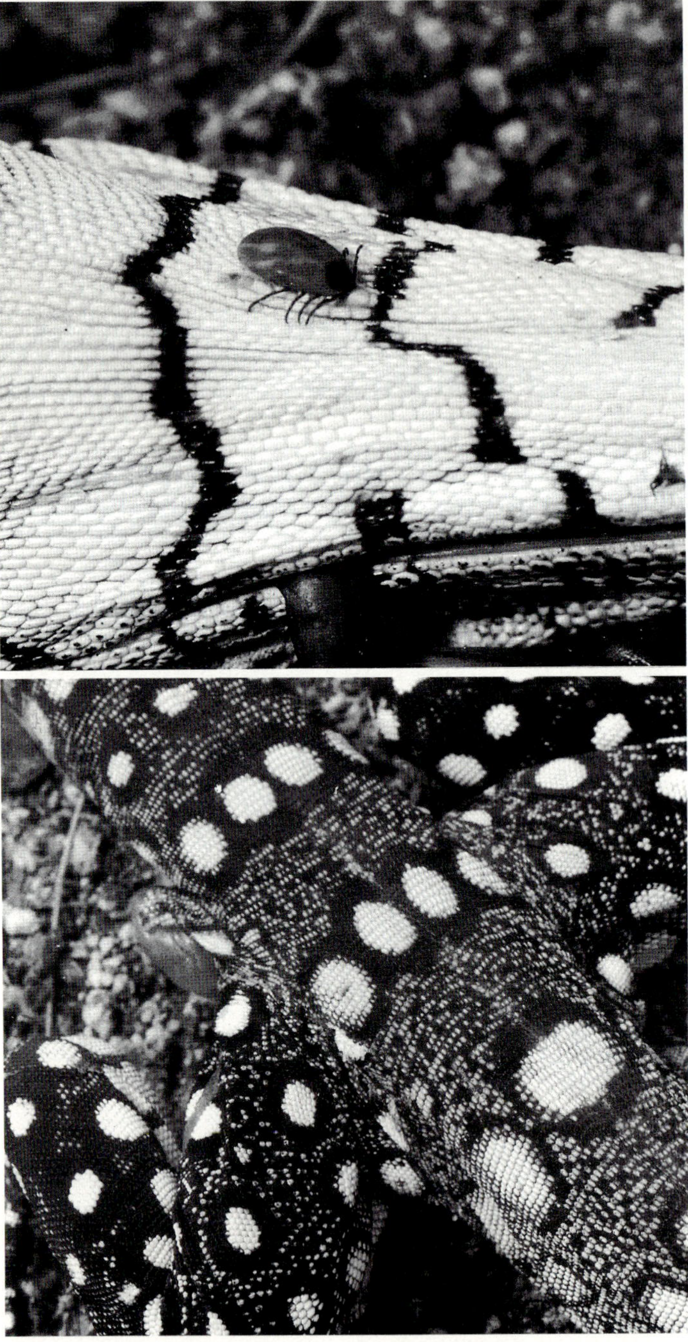

Fig. 23. Dorsal (*above*) and ventral (*below*) surfaces of juvenile perenti (*Varanus giganteus*) showing attached ticks (Ixodidae) (Central Australia)

Fig. 24. Larval ticks (*Rhipicephalus* spp.) awaiting chacma baboons and ohter hosts (Namib desert)

5.3 Endoparasites and their Transmission

Parasites have not been well studied in desert communities. Nevertheless, they represent an extra trophic level, and some of them are themselves the hosts of other parasites. According to Polis (1991b), of all the animals living in the Coachella Valley, California, only the parasites of scorpions, lizards and coyotes (*Canis latrans*; Gier et al. 1978) have been studied in any detail. Scorpions supported eight species of pterygosomid mites as well as nematodes (McCormick and Polis 1990), while many genera of spiders are known to harbour nematodes of the family Mermithidae (Poinar 1985). Telford (1971) identified the protozoan and helminth parasites in ten lizards, in addition to which there were a number of mite species, while Gier et al. (1978) discussed the parasites of the coyote. Ectoparasites included mange mites, ticks (seasonal) and fleas (*Pulex* spp.). An internal parasite, the tape worm *Taenia pisiformis*, occurred in 60–95% of all intestines. The chief intermediate hosts of this are cottontail rabbits (*Sylvilagus* spp.), and the infestation rate of the coyotes is correlated with the percentage of rabbits in their diet.

As elsewhere, endoparasites are often transmitted by arthropods in the desert. Trypanosomes are by far the most important protozoan diseases of camels, and are transmitted by blood-sucking flies such as

Lyperosia, *Stomoxys* and *Tabanus* spp. Traditional camel owners move their camel herds away from fly-infested areas seasonally, and change camping sites regularly to avoid the hatching of dung-breeding, blood-sucking flies (Wilson 1984). As far as human beings are concerned, the desert is a very healthy environment. Diseases transmitted by sandflies (*Phlebotomus* spp.) are, however, relatively common: the most imporant of these is leishmaniasis. The species of *Leishmania* (Trypanosomatidae) that are pathogenic to man can also utilize other hosts, such as rodents and jackals, to which they are not harmful (Dobson 1989). Two other zoonotic diseases transmitted by sandflies and common to human populations in arid regions are sandfly and oroya fevers (Cloudsley-Thompson 1977a). Sandgrouse (Pterocleidae) and sociable weavers (*Philetarius socius*) lack blood parasites (*Haemoproteus* and *Leucocylozoon* spp.: Haematozoa) in semi-arid regions of South Africa, probably on account of the scarcity of breeding habitats for the ornithophelic biting midges (Ceratopogonidae) and blackflies (Simuliidae), their commonest vectors (Little and Earlé 1995).

Where there is water there are parasitic diseases. Malaria has long been a menace in the oases of the Sahara. The malarial parasites (*Plasmodium* spp.: Sporozoa) exploit vertebrate hosts for asexual reproduction and then produce gametocytes, which are passed on by mosquitoes or sandflies in which the parasite undergoes both sexual and asexual reproduction. The geographical distribution and abundance of the parasites depends upon the sizes of the populations of both hosts, the precentage of vertebrates that are susceptible to malarial attack, the life span of the vector and its biting behaviour, as well as many other details of the natural history of both hosts. The western fence lizard *Sceloporus occidentalis* (Iguanidae) is host to the parasite *P. mexicanum*, whose invertebrate hosts are sandflies of the genus *Lutzomyia*. According to Schall and Marghoob (1995), the prevalence of malaria varies between 0–50% among the lizards in northern California. Males are infected more often than females, while larger and older individuals are infected more often than younger ones. The abundance of vectors is not related to the prevalence of malaria at a site. Sandflies are found in some sites where malaria is rare or absent, including those at higher altitudes. The pattern of distribution appears not to be random, but instead is a cycle of long duration (ca. 10 years).

In much of North Africa, up to 70 or 80% of the people are afflicted with trachoma, and many of them are left partially or completely blind as a result. Trachoma is spread by flies, the scourge of the oases, where they are much more common than in the surrounding desert. So, too, are the amoebae and bacteria responsible for different types of dysentery (Cloudsley-Thompson 1974a, 1977a), while the oasis water is home to snails (*Biomphalaria* and *Bulinus* spp.) that are intermediate hosts to the digenetic trematodes (*Schistosoma* spp.) responsible for bilharzia.

As already mentioned, seasonal timing may be a key factor in the transmission of parasites, and the monogenean trematodes that parasitise amphibians are a case in point. North American spadefoot toads (*Scaphiopus* spp.), for example, are inactive throughout most of the year. They avoid the harshest desert conditions by aestivating up to 1 m below the surface of the soil. Each year they emerge on the first night of heavy rain - generally in early July in Arizona - and spawn in the newly formed pools. The subsequent period of favourable conditions persists for little more than 8 weeks, and the actual season of activity is invariably much less. The toads are nocturnal and, during the rainy season, spend the day buried a few cm below ground, emerging to forage on nights following rainfall when the desert surface is relatively damp. This usually occurs less than 20 times in a season. In September, the toads retreat into their deep hibernation burrows, where they remain inactive for the next 9 or 10 months, surviving on the reserves of fat that were accumulated during their brief period of feeding (Tinsley 1990).

The elusive life style of the spadefoot toads imposes severe constraints on the transmission of their parasites. There are only limited opportunities, during the brief season of activity, for infective stages to be released by one toad and to be ingested by and develop within an invertebrate host, and for this to be eaten by another toad. The lack of digenetic trematodes is related to the absense of snails - obligatory intermediate hosts in their life cycles - and only one parasite is common, the monogenean *Pseudodiplorchis americanus*. This species infects more than 50% of *S. couchii* with a mean of about five adult worms per infected toad, while burdens of juvenile parasites may exceed 300 worms per host (Tinsley 1989).

P. americanus has a direct life cycle involving only one host, and transmission is effected by swimming ciliated infective oncomiracidia whose discharge is triggered very precisely during the spawning of the host. The total annual reproductive output is thus restricted to a period of only 7 h. About half the toads present lose their burden of parasites entirely at this time, but the remainder carry chronic infections throughout the period of hibernation of the host. Parasitic infection is pathogenic and creates extra stress during hibernation; nevertheless, extensive field studies have shown no correlation between mating success and the extent of parasitism (Tinsley 1990). Survival of the parasites is closely correlated with the life cycle of their hosts which, in turn, is dependent upon the annual desert rainfall. This finely tuned relationship between parasite and host provides evidence of a balance which has often been suspected in natural parasite populations, but has relatively rarely been documented.

Insects in which only the immature stages are parasitic are known as protelean parasites. Although their hosts are usually invertebrates, several species of Diptera make use of vertebrates as hosts for their larvae.

These are almost invariably endoparasitic, and their infestations are termed myiasis. The flies that cause myiasis in man include many arid zone species. They have been reviewed by M. T. James (1947), Zumpt (1965) and, more recently, by Hall and Smith (1993). According to Zumpt (1965), myiasis evolved from both saprophagy and blood-sucking habits. Larvae of the greenbottle (*Lucilia sericata*) usually feed on carcasses, but occasionally on purulent wounds, where they consume only dead tissues. Those of *L. cuprina*, on the other hand, feed on healthy as well as dead tissue. The final step along this evolutionary path has been taken by flies whose larvae rely entirely upon sores on vertebrates to supply their nutritional requirements. They include *Chrysomyia bezziana*, *Callitroga hominivorax* and *Wohlfahrtia magnifica*. Zumpt's (1965) postulated sanguinivorous origin of myiasis, like the saprophagous one, begins with a relatively unspecialised feeder in decomposing organic matter, such as *Muscina stabulans*, whose larvae can pierce the skin of a young bird or mammal in the nest of which they are living. This facultative ectoparasitism could then develop into an obligatory relationship, as is the case with the larvae of several other calliphorid species (see discussion in Askew 1971).

5.4 Parasitoids

The protelean parasites that attack invertebrates nearly always end up by killing them. They are known as parasitoids to distinguish them from typical parasites: the latter are much smaller than their hosts, which eventually, over a long period and many generations, become adapted to tolerate them (Askew 1971). Parasitoids are intermediate in behaviour between parasites and predators. At first, they operate as parasites, avoiding the vital organs of their hosts; but later, they behave like predators and consume the host which, therefore, cannot develop tolerance. Parasitoids commonly occur among Diptera and Hymenoptera, rarely in Coleoptera and Lepidoptera. They attack only in the larval stage and are often monophagous, whereas true predators attack in all developmental stages, and are seldom monophagous. Multiparasitism, the simultaneous infestation of the same individual host by two or more species of primary parasitoid, may cause the death of one or all of the parasitoides. Superparasitism is the name given to multiparasitism when two or more parasites of the species occur in the same host, while hyperparasitism occurs when the primary parasitoid is attacked by its own specific parasitoids. It seems probable that hyperparasitism is a development of multiparasitism, and it is indeed sometimes difficult to distinguish between the two.

Eggs, larvae and pupae are frequently attacked by parasitoids, but adults only infrequently. For instance, the eggs of desert grasshoppers are parasitised by Bombyliidae. Parasitoid and solitary habits sometimes converge. For example, the solitary Pompilidae and Tiphiidae both sting their prey and paralyse it before laying an egg on its body. Whereas pompilids place their spider prey in a subterranean chamber, however, the adult tiphid does not construct a burrow for the reception of the host. The Pompilidae are comparatively large wasps – indeed, tarantula hawks (*Pepsis* spp.) of the Sonoran desert are among the largest members of the order. After capturing and paralysing a spider with their stings, most species make their nests in burrows in the ground. The lower ends of these are enlarged for the reception of spiders, to each of which an egg is attached. The entrance to the burrow is then closed with earth and the young are left to themselves. Not all Pompilidae are digger wasps – some species make thimble-shaped nests of mud which are provisioned with spiders. The Thynninae is a subfamily of the Tiphiidae, most of whose members lay their eggs on the larvae of Scarabaeidae. The females are apterous and, in most species, are carried about by the males in a prolonged mating fight, during which they obtain food from flowers. The Scoliidae is an extensive family of large, hairy Hymenoptera whose prevailing colour is black (Sect. 3.8). The larvae are ectoparasites or parasitoids of larval Scarabaeidae or, rarely, of larval Curculionidae. Other important families that have adopted a parasitoidal mode of life are Ichneumonidae, Braconidae, Mutillidae, Trichogrammidae and Chalcidae. All but the last are strictly parasitic. The most important families of Diptera containing parasitoidal species are the Sarcophagidae and Tachinidae (Richards and Davies 1977).

Although spiders can do little except flee from their enemy, insects have evolved numerous morphological and behavioural defenses against parasitoids. Of these, structural defences, evasion and aggressive behaviour tend to be employed more than defensive secretions, symbiosis with ants or parental guarding. Most effective are combinations such as thrashing behaviour combined with long hairs or hard cuticles, gregariousness combined with defensive oral sections, or head-jerking combined with spines. Resistance increases with age: eggs are more vulnerable than larvae, larvae than pupae, pupae than adults. The subject has been reviewed by Gross (1993) and is no less relevant to the insects of arid regions than to those of more humid biomes.

Defenses used against predators are usually ineffective or less effective against parasitoids, but some are effective against both. For instance, the larvae of several species of Lepidoptera are stem-borers and feed inside the shoots of plants. Others are miners which exploit the underlying layers of plant leaves, while still others form galls. Such feeding specializations do not only result from competition with other herbivorous insects: the barrier of plant tissue surrounding them is a

protection from potential predators and parasitoids alike. Some parasitoids, such as ichneumon wasps, have evolved specializations for overcoming such barriers. Caterpillars also construct protective retreats, sometimes pulling together the leaves of plants and binding them in place with silk (Lederhouse 1990).

Some insects avoid parasitoid attack by habitat separation. For example, butterflies may locate their larval host plants, and then oviposit on dead vegetation or on the ground nearby; caterpillars may regularly commute between feeding sites and resting sites. Zebra swallowtail (*Eurytides marcellus*) caterpillars regularly rest away from pawpaw leaves when not feeding on them (Damman 1986). Many caterpillars pupate some distance from their host plants after completion of larval feeding, while others, such as the monarch (*Danaus plexippus*), evade high parasitoid mortality by migrating (Urquhart 1960). In this species, the frequency of parasitoids increases from generation to generation when the population remains in a particular area (Lederhouse 1990).

The role of parasitoids in the regulation of natural insect populations is uncertain, and much of the evidence for the importance of parasitoids as regulatory factors comes from the analysis of theoretical models. Long-term population studies in the field have not been carried out in desert areas and, even in regions where they have been, they have failed to resolve the issues at stake (Hassell and Godfray 1992). The behavioral and evolutionary ecology of parasitoids has been reviewed by Godfray (1994; see also Waage and Greathead 1986).

5.5 Kleptoparasitism and Slavery

Kleptoparasitism or piracy occurs when one species robs another of its food. This is frequent among most birds, but has developed as a regular habit in only a few. It is seen in a number of raptors, some of which inhabit arid regions, but seems to be particularly common among fish-eating species, such as the African fish-eagle (*Haliaetus vocifer*) whose particular victim is the osprey (*Pandion haliaetus*) although it also robs kites (Milvinae) and herons of their food. Mammalian carnivores, such as lions, hyaenas and jackals, frequently steal each other's kills and drive away the vultures, kites and marabou or adjutant storks (*Leptotilos* spp.) that attempt to rob them of their own. Kleptoparasitism is also found among spiders, large orb-web builders (*Nephila* spp.: Araneidae) in particular occasionally being robbed by smaller spiders such as *Argyrodes* spp. (Theridiidae; Lubin 1986). Social spiders (Sect. 3.9), likewise, are regularly exploited by kleptoparasites. The invading spiders not only steal prey, but also attack egg sacs, young and adults (Tietjen 1986).

Slave-making or dulosis is characterisitic of certain species of ants whose workers raid the nests of other species of ant and remove pupae. These are then reared as slaves in the nest of the dulotic ant species, and forage food for their captors. Many species of ants are known to carry back to their nests the bodies of enemies killed in territorial raids. These are later eaten. It is probable that territorial raids have led to slave-making in a few species, some of which inhabit arid regions. The evolution of social parasitism has been discussed in detail by Hölldobler and Wilson (1990).

5.6 Social Parasitism, Commensalism and Mutualism

Termites abound in the tropics, including the desert regions, and their colonies are frequently inhabited by social parasites or inquilines belonging not only to different orders of insects, but also to other classes of arthropods. The relationships between these inquilines and their hosts are very similar to and parallel those between mymecophilous arthropods and ants. They range from true guests and indifferently tolerated lodgers to scavengers and predators. Social parasitism in termites and ants is complicated. Some species of small ants build nests near those of larger species and either feed on the refuse of the latter or rob the host workers when they return from foraging. This is an example of klepto-parasitism (Sect. 5.5). Social parasites are more common, however, in temperate regions than they are in the tropics or in subtropical deserts (E. O. Wilson 1971).

Numerous species of bugs, beetles, flies and other insects, as well as of woodlice, millipedes, spiders, lizards and so on, are to be found living in the nests of ants in different parts of the world (Hölldobler and Wilson 1990). Some are commensal (see below) and feed on debris in the nest, some are social parasites, but others are synoeketes, spending part or all of their lives as indifferently tolerated lodgers. *Metopina pachycondylae* (Phoridae) lives in colonies of the ponerine ant *Pachycondyla harpax* in Texas. Its small larvae cling to the necks of ant larvae by means of suckers, encircling their hosts like collars. Whenever the ant larvae are fed, the larvae of *M. pachycondylae* uncoil themselves and partake of the feast. Sometimes, when there is no food within reach, the phorid larva will tweak with its mandibles the skin of its host, or that of a neighboring ant larva, whose wriggling then incites the worker ants to bring a fresh supply of food (see discussion in Cloudsley-Thompson 1965).

The nests of invertebrates, as well as those of vertebrates, provide shelter for many small, itinerant arthropods and a source of food for scavengers and parasites (Sect. 5.2). For instance, the larvae of many species of desert Carabidae, such as *Thermophilum* and *Graphipterus* spp., develop in ants' nests containing broods upon which they feed. If the ants of a plundered brood chamber encounter a beetle larva, they attack it and the larva escapes by burrowing into the surrounding sand (Paarmann 1979, 1985). Adaptations of these beetles include an unusually large number of small eggs and small emerging larvae which escape the rigours of the dry season in the desert as a result of commensalism with ants. Despite the carnivorous habits of their hosts, the nests of spiders, especially social species, are habitats for complex communities. Nest associates of *Stegodyphus mimosarum* and *S. deserticola* (= *Magunia dumicola*; Sect. 3.9) comprise five classes: (1) predators and parasitoids, (2) kleptoparasites, which steal or share prey items caught in the webs, (3) scavengers, which feed on nest debris, (4) spiders, insects, and occasionally small mammals, which use a portion of the nest as a retreat, (5) predators and parasites of the arthropods in the first four classes (Meikle Griswold 1986), Downes (1994) adopted the same system of classification in his study of the nest associates of the Australian *Badumna* (= *Phryganoporus*) *candida* (Desidae), a species of moist, deciduous forest. Henschel (1991) has recently found that solitary specimens of *S. deserticola* sometimes occur, and he suggested that environmental conditions occasionally permit individual spiders to escape from the competition that occurs within the social unit. Solitary individuals occupy small nests that they build by themselves at greater heights up *Acacia* trees than are the nests of the large social units in the area, which contain as many as 25–50 spiders.

When two animals live in a close and harmonious association with one another from which only one derives benefit, the relationship is known as commensalism. Sometimes more than one species of termite may inhabit the same termitarium as commensals. Usually the smaller and weaker of two commensal species is mildly kleptoparasitic, but steals so little food that its partner is not noticeably affected. Small muscid flies sometimes accompany large, blood-sucking horseflies (Tabanidae) and lap up some of the blood from the victim's wound, thereby obtaining an advantage that it does not reciprocate. It is but a small step from this to thieving, a mild form of parasitism by no means unknown amongst the Diptera. An example is afforded by *Bengalia depressa*, a marauding muscid fly which exhibits considerable skill in snatching its prey from the jaws of army ant workers in East Africa (Thorpe 1942). *Bengalia* spp. not only feed as larvae in the nests of termites but, as adults, hover over columns of ants on the march and snatch from them eggs, larvae, pupae, or even adult ants themselves (Sect. 2.2). Not only can commensalism lead to predation, as this example shows, but it may also be a stage in the evolution of ectoparasitism (Sec. 5.6).

The utilization of another creature for conveyance only is perhaps a particular type of commensalism and, to this, the name phoresy is commonly applied. An example is afforded by the borborid fly *Limosina sacra*, constantly to be found riding on the dung beetle *Ateuchus laticollis* in Algeria – apparently for the purpose of ovipositing in the particular choice of faecal matter that the beetle has collected and consolidated for its own use. (Borboridae often visit animal droppings in large swarms and lay their eggs there.) *L. sacra* or a related species is likewise frequently clustered on *Copris hispanus* in southern Tunisia. Though perfectly able to fly, *L. sacra* refuses to be dislodged by the antics of the beetle, though it readily evades capture on its own account by flight. In a similar way, Mallophaga are frequently transported from one host to another by louse flies (see Cloudsley-Thompson 1965).

When two organisms live in an association from which both partners benefit, the relationship is known as mutualism or symbiosis. In general, symbiosis is a much closer association than commensalism, but the demarcation line between the two is often hard to draw. A familiar example is provided by lichens, whose success in desert regions, as elsewhere, depends upon the fact that they are a symbiotic association between an alga and a fungus, neither of which could probably survive on its own. Lichens are able to absorb moisture from unsaturated air as well as from fog, and can withstand any amount of desiccation. They recover immediately after they have been wetted (Fig. 25). Lichens are particularly conspicuous in the cool coastal Namib desert, where they display a wide range of colour and form. Some are fructose, some foliose, and their colours range between orange or green to grey and black. Most are firmly attached to stones and the gypsum soil, but one species, *Xanthoaculina convoluta*, is completely unattached. It is blown about by the wind, collects in small depressions, and becomes entangled among other lichens (Seely 1987).

Another well known example of mutualism is afforded by termites and their associated gut faunas of flagellate Protozoa, without which cellulose could not be digested. Termites whose symbiotic Protozoa have been removed experimentally lose weight and eventually die if they are prevented from acquiring new symbionts. There is also a very close association between *Macrotermes* spp. (Macrotermitini) and the fungus gardens in their nests. The chambers in which the fungi grow are scattered throughout the termite nest, sometimes being concentrated around the royal cell. Two genera of fungi occur in termite fungus gardens; *Xylaria* (Ascomycetes) and *Termitomyces* (Basidiomycetes). The former does not produce fructifications in inhabited nests and is not confined to termitaria, but the latter is exclusively termitophilous and produces small white spheres which are eaten by nymphs and workers (E. O. Wilson 1971).

Symbiotic relationships have not evolved among desert organisms in direct response to the environmental extremes experienced in such places. Nevertheless, they do enable certain taxa to inhabit arid environments in which they could not survive on their own.

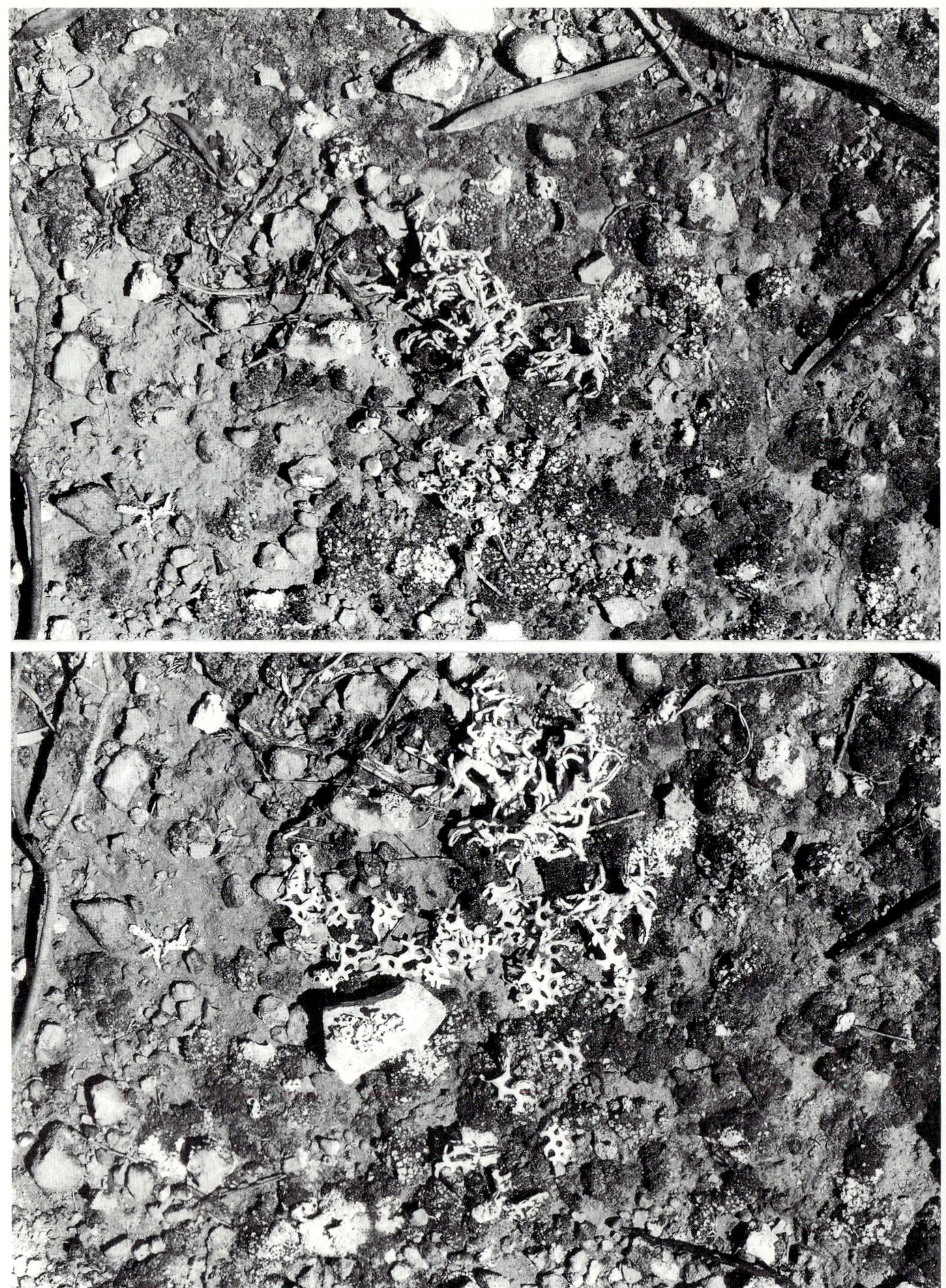

Fig. 25. Desert lichens in South Australia. *Above* Desiccated; *Below* immediately after being wetted

5.7 Evolutionary Trends in Parasitic Relationships

Parasitism has evolved over and over again in different taxa and in different ecobiomes. As a matter of convenience, parasitic relationships have been discussed here under the categories of ectoparasites (Sect. 5.2) and endoparasites (Sect. 5.3), along with parasitoids (Sect. 5.4), social parasites, commensals and mutualists (Sect. 5.6). This arrangement is, however, rather artificial, for innumerable intermediate stages exist. Furthermore, the attentions of ectoparasites may engender reactions in the internal organs of the host so that, when these interactions are considered, the distinction between ectoparasitism and endoparasitism becomes blurred. Comparable ramifications have been noted with the other categories. Parasitism can be regarded as a special type of predation. Instead of killing and devouring its prey at one time, the parasite, by virtue of its small size, lives on or in its host, eating it or its food, little by little. Parasitoids start off as parasites but, when they grow larger, they become more like internal predators and devour the internal organs of their host, thus finally killing it. Commensals may evolve into ectoparasites via the blood sucking route, while endoparasites could, perhaps, become symbiotic through the development of tolerance between them and their hosts (see discussion in Cloudsley-Thompson 1965).

As far as the parasites of desert animals are concerned, some modifications of life history may be essential for their transmission (Sect. 5.3). Outside the range of tsetse flies, the only trypanosomes able to survive are those that are transmitted either venereally, or mechanically by the bites of horseflies and clegs (Tabanidae). An example of the latter is *Trypanosoma evansi*, the causative agent for the fatal disease surra in camels in North Africa, which is known also to parasitise horses, donkeys, cattle and dogs, while *T. equiperdum*, the causative agent of the chronic disease dourine of horses and donkeys throughout the world, does not require as vector but is transmitted sexually. It can, however, also be transmitted directly by horseflies and stable flies (*Stomoxys* spp.). None of these passes through developmental stages in its invertebrate vector (Cheng 1986).

Because individuals vary, some degree of pre-adaptation must exist in all organisms. Consequently, some are more likely than others to adopt blood-sucking or commensal habits. Others, such as the larvae of higher flies (Calliphoridae and Muscidae), which normally feed on decaying flesh, already possess adaptations that may allow them to survive in the alimentary canals of vertebrates that inadvertently swallow them. In such cases, the transition from a free-living existence to endoparasitism can be direct. Parasitic existence during the larval stages of an animal's life cycle is not as restrictive as it is in the adult stages. It is prob able that both the social Hymenoptera and the Cyclorrhapha may likewise have descended from ancestors all of which had parasitic larvae, although, at the present time, only a small proportion have retained this habit (Rothschild and Clay 1952).

The low population densities of potential hosts in arid regions makes deserts rather unsuitable habitats for parasites, which, in general, require a large host population density to sustain a viable infection. The number of hosts required for this purpose tends to vary inversely with the pathogenicity of the parasites, and those found in deserts are usually relatively benign. Indirect life cycles seem less common in parasites from arid regions, probably because heteroxenic cycles require at least one of the hosts to occur at relatively high density. Concentration of hosts in oases or at the time of rain increases selection pressure on efficiency of transmission which, as in the case of spadefoot toads (Sect. 5.3), leads to the synchrony of the reproductive activity of the parasite with that of its host. Behavioral adaptations of parasites tend to have a strong physiological component in deserts because climatic conditions are so harsh (Dobson 1989).

6 Plants and Herbivorous Animals

Plants have a variety of defensive mechanisms which help to reduce damage from herbivores. These may be physical, chemical, physiological or biotic: they vary both quantitatively and qualitatively among and within species. Many plants also show daily, seasonal and annual fluctuations in resistance, both within individuals, as well as in ontogenetic changes. Synchrony between the quality of plant food and peak feeding on it by herbivores is often critical in determining whether the abundance of herbivores increases or decreases. In arid regions, outbursts of plant growth result from irregular precipitation. Vertebrates usually build up metabolic food reserves to withstand lean periods, while invertebrates enter into diapause.

Most current work on the anti-herbivore systems of plants assumes a cost, associated with the production and maintenance of a given level of defence against both invertebrate pests, mainly insects, and browsing and grazing vertebrates (Rhoades 1979; Gulmon and Mooney 1986). Fagerström (1989) proposed that this cost should be expressed as a loss in the relative growth rate of somatic tissue. This applies not only to defensive chemicals (Sect. 6.2), but also to systems such as thorns and spines (Sect. 6.3), as well as to other kinds of costly activity in plants, including the production of nectar or of advertising structures (Sects. 6.1, 7.4). The allocation of resources to reproduction and defence has been reviewed by Bazzaz et al. (1987). It should also be remembered that the pressure of insect pests is heavier in tropical and subtropical climates than it is at higher latitudes, because the winters are less severe there, and do not set back insect pests to such a degree.

Re-growth is a generalized response by plants to all types of damage, not only that caused by herbivores, and rapid re-growth may have been selected by any one of a large number of factors or, more probably, by a combination of all of them. According to Belsky et al. (1993), there is no reason to assume that the increases in productivity which occasionally follow herbivory constitute a specific evolutionary response to herbivory, as has been suggested in the past by a number of authors including J. Owen (1980).

Indeed, there is no evolutionary justification and little actual evidence to support the idea that plant-herbivore mutualisms should be likely to evolve. None of the assumptions that underlie discussions of the benefits of herbivory is necessarily correct, but there is no compelling evidence that herbivory does increase plant fitness. On the contrary, it is often highly deleterious, sometimes less so, but generally harmful, nevertheless (Crawley 1983).

The term predator is not infrequently applied to animals that eat plants or their seeds (e.g. by R. M. M. Crawford 1989), but this usage has been criticised by Heatwole (1995). I, also, have not used it in that context for two reasons. First, it may be confusing to equate a herbivore with an animal that preys on other animals and, secondly, because, as D. F. Owen (1990) and others have pointed out, using the word predator to describe the eating of plants by animals immediately suggests that the action is detrimental to plants. This implication is not only unnecessary, but could bias subsequent interpretation if elaborated. It has even been suggested that plants may actually benefit from being browsed or grazed, although this has been disputed, as mentioned above. Moreover, seeds often germinate better after being ingested (Gutterman 1993) and are usually well dispersed when the fruits or pods have been eaten by birds or mammals (Sect. 7.5).

Despite the fact that they provide the basic source of food for all animal and microbial life, plants are not usually consumed to the point at which they are no longer able to support the organisms that depend upon them. Much ecological research has been undertaken on the mechanisms that determine the balance between maintaining population numbers and not over-exploiting the resources of the habitat. Nevertheless, plants are sometimes severely affected by plant-eating animals, both directly and indirectly. Thus, seed production is extremely sensitive to attack on the leaves by herbivores, and reduced seed production is typically the first response of a plant to defoliation (Crawley 1983).

Many desert plants owe their unpalatability to secondary metabolites (R. M. M. Crawford 1989; S. D. Smith et al. 1996), but no defence, as we have seen, can ever be perfect, and a balance is maintained through the buffering effect of numerous interacting variables. For example, dorcas gazelles (*Gazella dorcas*) in the Negav desert sand dunes feed on a single plant species, the madonna lily (*Pancratium sickenbergeri*: Amaryllidaceae), and dig holes to remove the subsoil stems and bulbs. A negative correlation between the size of the plant and the amount eaten is due to the increased probability of reaching the bulbs of smaller plants. Consequently, gazelles exert a selection on lilies to grow at sufficient depth to avoid damage to their bulbs. Conversely, by selecting plants with their bulbs near to the surface, the gazelles maximise food intake and minimise the cost, in energy, of digging. After rain, only the tips of emerging leaves are eaten, plants with more numerous and larger leaves being selected (D. Ward and Saltz 1994).

Individual plants may vary in their resistance to browsing, and this complicates the situation. For instance, creosote bushes (*Larrea triden-*

tata: Zygophyllaceae) vary considerably in their responses to browsing by jackrabbits (*Lepus californicus*). Some individual plants tend to be only lightly browsed, while others are fed upon several times each year. The latter are more likely to be browsed again than are individuals with a history of light browsing and which have high constitutive resistance. The innate degree of resistance can, however, be increased by exposure to herbivory, and this induced resistance seems to be stronger in plants with high constitutive resistance (Ernest 1994).

In summarizing population studies of the saguaro cactus (*Carnegia gigantea*; Fig. 7), C. J. Krebs (1985) pointed out that the saguaro may live for 175 years or more, and its populations turn over slowly. In 1910, Shreve concluded that the species was not maintaining itself and was dying out. Steenbergh and Lowe (1977), however, pointed out that the reproductive potential of the saguaro is enormous. Young plants reach a height of 2–3 m in 30–40 years. They flower and fruit annually, producing an average of 200 fruits, containing about 400 000 viable seeds. Given the density of mature plants in the Saguaro National Monument, some 40–60 million seeds are produced each year. Of these, only 4 per 1000 survive to become seedlings, resulting in 160 000–240 000 seedlings per ha or 16–24 per m^2, varying according to summer rainfall. Many of these are eaten or dug up, especially by ground squirrels. Of the rodents, only the woodrat (*Neotoma albigula*) can subsist on a diet of saguaro, because the cactus contains secondary compounds in the form of oxalates that are poisonous to most other mammals. Many rodents eat saguaro in small amounts for its water content, but not for its nutritional value.

Other losses are caused by drought during the first few years of life. Small saguaros are often found beneath desert trees and shrubs, but not in the intervening spaces. They become independent of their nurse plants at a later stage, when their water storage capacity begins to develop. The decline of saguaro reproduction in the 1870s and 1880s was associated with rapid growth of the cattle industry – resulting in heavy grazing, trampling and the removal of nurse trees and shrubs. More important than any of these, however, is freezing weather. Saguaros do not freeze at 0 °C because they supercool to between –3 °C and –12 °C. This is their lethal limit, and larger plants are more susceptible to freezing than smaller plants. Population changes are therefore determined mainly by winter weather in association with drought, with being eaten or trampled, and with the survival of nurse plants until the cacti beneath them have become firmly established.

In general, herbivore population cycles tend to be engendered by plant-herbivore interactions, rather than by reactions with predators, parasites and disease, or the responses of plants to their environments (see Fritz and Simms 1992). In arid regions, however, the pulse provided by precipitation is of paramount importance.

6.1 Vegetative Crypsis, Mimicry and Deception

Herbivory – browsing and grazing – is probably more important to the scattered vegetation of arid regions than it is to the denser vegetation in mesic ecobiomes. Some desert plants, heavily grazed elsewhere, are often restricted to jebels, or the sides of wadis, where they are less accessible to camels, sheep and goats. Examples from the northern Sudan include: *Abutilon fruticosum, Hibiscus micranthus* and *Gossypium anomalum* (Malvaceae), *Forsskaolea tenacissima* (Urticaceae), *Heliotropium strigosum* (Boraginaceae), *Cleaome scaposa* (Capparaceae), *Cenchrus ciliaris* (Graminae) and *Blepharis ciliaris edulis* (Acanthaceae) (Fig. 26; Cloudsley-Thompson 1968).

6.1.1 Protective Resemblance to Stones and Sticks

Another way of avoiding the attention of mammalian herbivores is by means of crypsis. Some succulents produce sticky, succulent trichomes on their surfaces which trap dust and sand particles so that the plants that bear them strongly resemble their substrates. For instance, *Crassula*

Fig. 26. *Blepharis ciliaris edulis*, a plant often restricted to mountainous jebels in the Sudan because it is heavily grazed elsewhere by camels. (Cloudsley-Thompson 1968)

alpestris (Crassulaceae) in South Africa, North American cacti (e.g. *Ariocarpus kotschoubeyanus, Pediocactus sileri* and *Sclerocactus spinosior*), and *Psammophora* spp. (Aizoaceae) may accumulate sufficient sand to make them resemble low mounds (Wiens 1982).

Mimicry is an evolutionary concept, termed adaptive resemblance by Starrett (1993). As in animals (Sect. 3.5), protective resemblance to stones is not unusual among desert plants. The phenomenon is extensively developed in the arid regions of southern Africa, and is also to be encountered in Somalia, Ethiopia and North America: numerous genera and species are involved, especially among the Aizoaceae. Protective resemblance to stones also occurs in North American species of cacti, including *Ariocarpus, Pediocactus* and *Sclerocactus* spp. Most *Ariocarpus* spp. resemble large, angular rocks, while *P. peeblesiana* and *S. pubispinus* look more like pebbles or small stones (Wiens 1982). Stone mimicry has clearly evolved independently on many occasions and has taken numerous forms. The plants concerned usually do not, however, simply resemble stones. They duplicate the shape, size, colour, texture and fracture lines of the particular rock substrate to which they are restricted. Sometimes the entire plant is buried beneath the surface of the soil, with only the tips of the leaves exposed, as in *Lithops salicola* (Fig. 27), *Nananthus vittatus*, and *Conophytum* spp. (Aizoaceae) of the Kalahari, and certain soil cacti of the northern Atacama desert of Chile (Walter and Stadelmann 1974). Although the advantages of dwelling beneath the soil surface may lie mostly in the reduction of water loss through transpiration (R. M. M. Crawford 1989), avoidance of the atten-

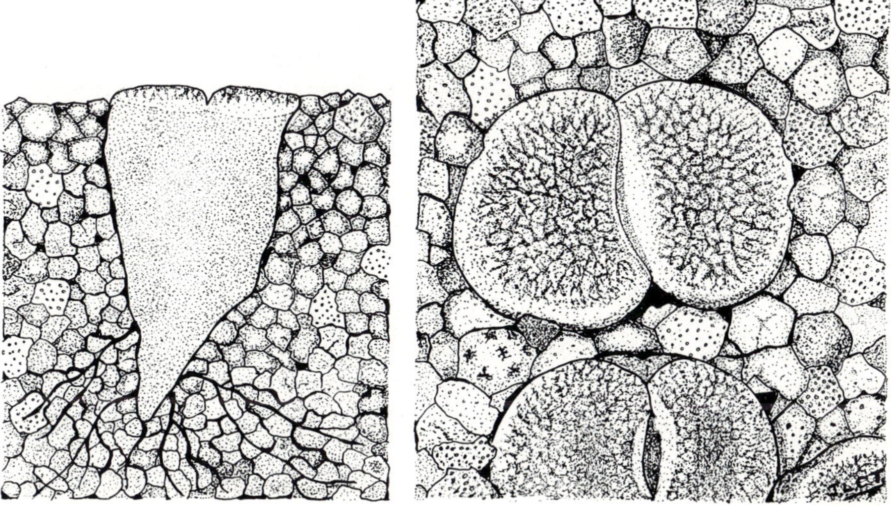

Fig. 27. *Lithops salicola. Left* Diagram of growth form; *right* close-up view of expanded leaves

tions of grazing herbivores may also be beneficial, and the resemblance to stones is certainly not fortuitous.

Members of six families of plants in southern and eastern Africa include leafless succulents whose wrinkled, greyish stems grow typically among the weathered grey branches of dead shrubs. Examples from the Karoo region include *Crassula lycopodioides* (Crassulaceae), *Euphorbia* spp. (Euphorbiaceae), *Orthonna protecta* (Compositae), *Piaranthus parvulus* and *P. punctatus* (Asclepiadaceae) which grow in association with *Ruschia* spp. (especially *R. spinosa*: Aizoaeceae) – semi-palatable ecological dominants and ideal associates for stick mimics. *Rhytidocaulon* spp. (Asclepiadaceae) are also stick mimics. *R. subscandens* is eaten both by animals and by Somali tribesmen, who consider it to be sweet and fleshy. They devour it raw and cooked. It grows hidden in thorny bushes such as *Balanites aegyptiaca* (Balanitaceae), while *R. macrolobum* grows in south-western Arabia among dead sticks, where it is beautifully camouflaged. *Euphorbia catervifolia* of the Karoo appears to grow singly or away from dead shrubs, while small *Anacampseros* spp. (Portulacaceae) resemble fragments of cushion plants (*Cheiridopsis* spp.: Aizoaceae). Larger species (e.g. *Anacampseros papyracea*) resemble the faecal droppings of birds, and are known by the Afrikaaner farmers of Namaqualand as "goose droppings". Other smaller species have features in common with the pellet-like faeces of ungulates, as does *Crassula alstonii* (Wiens 1982). The dry fruits of *B. aegyptiaca* resemble camel dung sufficiently closely for real dung to be tasted (but not eaten) by sheep and goats, while the plate-like mats of *Anthobryum traindra* (Frankeniaceae) in the high Andes are closely reminiscent of cow pats (G. E. Wickens, pers. comm.). What was their original model?

Camouflage is associated with two life-forms in flowering plants: succulence and aerial parasitism (see below). Many succulents, especially cacti, ice plants (Aizoaceae) and mistletoes are highly palatable, and possess few phytotoxins. The succulent habit is largely associated with aridity, and succulents may be even more important to herbivores as sources of water than of nutrients. Some cacti appear to mimic dead grasses. The central spines of these species are much enlarged, resembling, in shape and colour, the dry, dead leaves of grass. In the south-western United States, *Pediocactus papyracanthus*, known locally as gramma grass cactus grows in or near fairy rings of the blue gramma grass (*Bouteloua gracilis*). Here it is unnoticed because the spines resemble the dried leaves of the grass (Benson 1969). *P. toumeyi* and *Sclerocactus wrightii* are further examples: not only gramma grasses, but *Hilaria* spp. also serve as models. *Opuntia glomerata*, which occurs on dry hills in western Argentina, has similar leaves, as does the Mexican *Coryphantha bumamma* and possibly *Echinocactus* spp. (Wiens 1982).

6.1.2 Mimicry in Mistletoes

At the present state of knowledge, crypsis appears to be the primary form of deception by which plants escape the attentions of herbivorous animals. Crypsis, however, must be dependent upon size – trees and shrubs can scarcely become cryptic. According to Wiens (1982), the shrubby loranthaceous mistletoes are probably the largest cryptic or, rather, mimetic plants; but they are unusual because of their spatial isolation as aerial parasites. Approximately 78% of Australian mistletoes such as *Amyema maidenii* and *Lysiana subfalcata* (Loranthaceae; excluding rainforest species) mimic their host plants to some extent, according to Barlow and Wiens (1977). They show three different leaf classes: (1) flattened, linear to lancolate types, common in *Eucalyptus* spp. and the phyllodinous acacias; (2) thick, rounded leaves, typical of mangroves and a few *Eucalyptus* spp. with neotenic leaves, and (3) linear, often compressed or rounded types, especially common in *Casuarina* spp., but also found on some *Acacia, Grevillea* and *Eremophilia* spp. It seems probable that selection has been achieved by leaf-eating possums (Phalangerinae), which use visual cues in foraging.

Alternatively, lycaenid and pierid butterflies could be the primary selective agents (although insects normally use chemical cues to locate their host plants, as already stated). Some African mistletoes, including *Tapinanthus kayseri, T. landertziae* (Loranthaceae) and *Viscum* spp, are cryptic, as are *Phoradendron* spp. (Viscaceae) in North America (Wiens 1982).

Mullerian mimicry is a long-established concept referring to situations in which two or more unpalatable, aposematic species resemble one another. Theoretically, this has the effect not only of sharing the loss incurred while predators are being conditioned to avoid them, but it also reinforces the effect so that predators learn more quickly (Cott 1940). Despite its widespread use, there are logical difficulties in the concept of mullerian mimicry, since the animals that exhibit it not only possess aposematic characters but are themselves all relatively unpalatable or obnoxious. Consequently, it might be argued that the system has no mimics, only models and "operators" (predators), and therefore it is really an example of group convergent evolution. Mullerian convergence must, nevertheless, be considered as a significant factor in the evolution of plant pollination systems (Wiens 1982; Sect. 7.4).

That protective resemblance and mimicry should occur in plants is not a new idea, but the information available is extraordinarily scattered and largely anecdotal. In addition to stone mimics and the mimics of sticks and faeces (Sect. 6.1.1), there are also some early references to host-parasite resemblances in mistletoes (see above). The subject was not considered seriously, however, until it was reviewed by Wiens (1982), who pointed out that mimicry evolves as a consequence of co-evolution. The major evolutionary development of mimicry in animals and plants has originated from diametrically opposed biological interactions – animal mimicry from antagonistic co-evolutionary relationships, and plant mimicry from mutualistic co-evolutionary interactions. This functional role reversal can be explained as follows: plants are largely protected from herbivorous animals by repellent or toxic chemicals (Sect. 6.2), and herbivorous animals, especially insects, locate their plant hosts primarily by using chemical cues. Conversely, animals, especially insects, are themselves largely subjected to predation by vertebrates that use visual search images to locate prey. Thus, crypsis and batesian mimicry are effective responses to visually oriented predatory animals, but not to chemically oriented herbivores. It is not surprising that, in exposed arid regions, mimicry among plants is more common and better understood than it is in less exposed, mesic environments where the vegetation is very much more luxurious.

6.2 Chemical Deterrence

We have seen above (Sect. 6.1) that plants, being both sessile and unable to employ crypsis, obtain protection from herbivorous animals mainly through the secretion of secondary repellent and toxic chemicals. Secondary compounds can be defined as those which do not function directly in the primary biochemical activities that support growth, development and reproduction of the organism in which they occur. In the case of plants, they are of importance in interactions between plants and animals and between plants and micro-organisms, as well as between plants and other plants. No less than morphological features, they are subject to the selective pressures of the environment. Probably millions of secondary compounds have been synthesised by plants in the course of their evolution, and those that increased the competitive fitness of the plant in which they arose have survived. Thus, the synthesis of 10% canavanine in plant seeds can protect them against most seed-eating beetles (Bruchidae; Sect. 6.4). Consequently, the expenditure of resources and energy in providing the enzymes for canavanine metabolism are a sound economic investment, costing the plant less than it would do to produce additional seeds to compensate for the losses inflicted by beetles in the absence of canavanine (E. A. Bell 1981).

The greater the number of species sharing the same environment, the more complex do their interactions become. Tropical rainforests are the most complex of terrestrial ecosystems, while deserts and polar regions are among the simplest. The co-existence of plants and animals in arid regions is possible only because herbivores there accept a wide range of food plants and are not host-specific as are many forest species.

Host specificity is possibly greatest among insects in both ecobiomes, although it is probably less so in deserts. At the same time, whatever their food plants, the nutritional demands of herbivorous animals in respect of amino acids and carbohydrates are remarkably similar. Some, however, need concentrated and nutritionally rich foods such as seeds and fruits, some live on nectar or sap while yet others, such as leaf-eating insects and mammalian ruminants, are able to digest larger quantities of roughage - often with the assistance of symbiotic microbes in the alimentary canal (Sect. 5.6; see R. M. M. Crawford 1989). The nitrogen content of the vegetation is a limiting nutritional factor for many herbivorous animals. Those that are adapted to feed on plants with low nitrogen contents have low growth rates and long life cycles. Plants that are rich in nitrogen suffer the severest attack not merely because they are nutritionally rich but because they can support insect species with rapid growth and short life cycles.

Many desert plants, like animals, are ephemeral and have short life cycles that are triggered by precipitation (Sect 6.7). They probably survive partly because their seeds germinate so rapidly, grow to maturity, flower, and produce new seeds within the span of a few weeks or even days. Their lives end before herbivores can reach them or, in the case of insects, reproduce in sufficient numbers in time to make serious inroads into their populations. The biomass of predatory arthropods does not respond as rapidly to increased productivity after rainfall as does that of herbivorous insects (Ludwig and Whitford 1981), and even the latter cannot keep up with the vegetation. Perennial desert plants are endangered to a greater extent by herbivorous animals, and it is among them that the widest variety of chemical deterrents is to be found. The resistance of plants to herbivores and pathogens is reviewed from several different aspects in Fritz and Simms (1992), while the interactions of herbivores with secondary plant metabolites are discussed in great detail by Rosenthal and Berenbaum (1991, 1992).

6.2.1 Secondary Metabolites and their Cost

The secondary metabolites that make plants unpalatable, toxic or repellent to herbivorous animals include: (1) alkaloids, (2) toxic amino acids, (3) toxic amines and peptides, (4) toxic proteins, (5) cyanogenic glycosides and nitro-glycosides, (6) coumarin glycosides, (7) steroid and triterpenoid glycosides, (8) irritant oils, (9) organic acids, (10) phenolic acids and tannins and (11) volatile oils. Their chemistry has been discussed by R. M. M. Crawford (1989) and in Rosenthal and Berenbaum (1991). The leaves of creosote bushes have a pungent smell, euphorbias contain latex, while the pods of *Cassia italica* (Leguminosae - Caesalpinioidea) contain a strong purgative. Whether or not the function of this is to make them repellent to herbivores is discussed below (Sect. 7.5). Tannins probably interfere with digestion, but this does not appear to have been demonstrated yet. Plants such as *Euphorbia* spp. (Euphorbiaceae) and the Asclepiadaceae contain poisonous or irritant latex. The Sodom apple (*Calotropis procera*: Asclepiadaceae) of the Sahara and Sahel is eaten by the grasshopper *Poeciloceros hieroglyphicus*, which sequesters its poisons (Sect. 7.3), but is avoided by other insects, goats and camels. Terpenoids and other aromatic compounds are produced seasonally by *Commiphora* spp. (Burseraceae) and other desert shrubs at the time of the rains. These are responsible for the characteristic scents of frankincense and myrrh, and deter plant-eating insects. The mesocarp of the fruit of *Balanites aegyptica* (Balanitaceae) is lethal to the snail vectors of schistosomiasis (*Biomphalaria glabrata* and *Bulerius globosus*) as well as to mosquito larvae and, therefore, presumably is harmful to other insects, as are the leaves of the neem or margosa tree (*Azadirachta indica*: Meliaceae; see review by Kloos and McCullough 1987).

In the past, the poisons produced by plants have been regarded only as anti-herbivory agents. It is now recognized, for instance, that fruit laxatives may also serve to cause seeds to be voided rapidly, which enhances their chances of germination (Murray et al. 1994; Sect. 7.5). Again, the discovery in recent years of anti-feedants and repellents further highlights the dynamics of co-evolution among plants and insects. Studies of insect endocrinology are now unravelling the details of the ways in which hormones regulate the lives of insects. It is, therefore, not surprising that plants should have evolved subtle defensive strategies focussed on the disruption of insect endocrine systems (W. S. Bowers 1991). Some plants are capable of synthesising animal hormones in their defence against herbivores. Not only are the reproductive hormones of birds and mammals mimicked, but also the insect juvenile hormone. Other plants produce substances that inactivate proteolytic and other insect enzymes (Crowson 1981). Hormonal defences of plants have been reviewed by R. M. M. Crawford (1989) and W. S. Bowers (1991).

The cost of defensive compounds, in terms of reduction in growth, due to the allocation of carbon and energy for their production, varies according to the rate of photosynthesis and the fraction of leafage allocated for this purpose. For this reason, qualitative defences are found only in plants with lower photosynthetic rates or leaf allocation fractions. When diverted from growth and metabolic functions to defence, the cost in terms of nitrogen is highest in plants with low nitrogen contents, as well as low photosynthetic rates and leaf allocation fractions. Consequently, nitrogenous defence compounds are found only in plants with high nitrogen contents. Such plants, however, are most sensitive to the carbon costs of defence and therefore produce defensive compounds in smaller quantities unless they have other secondary functions

(Gulman and Mooney 1986). In other words, although the metabolic cost of different defensive substances does not vary much except, possibly, in the case of tannins, the type of defence used depends upon the life history of the plant concerned.

Slow-growing plants with long lifetimes use more expensive, but immobile, compounds with low rates of turnover. As a consequence, the type of defensive substance used is affected by the C/N balance of the plant. Where soil nutrients are low, it is cheaper to produce carbon-based defence and vice versa. Even with higher concentrations of available nitrogen, it is more expensive to produce tannins (and mechanical defences – Sect. 6.3) than other chemical defensive substances – especially when the plant is under carbon stress. Even in carbon-rich environments, however, tannins (and mechanical defences) are more expensive than other secondary metabolites (Skogsmyr and Fagerström 1992). The stable co-existance of different defence strategies may be maintained by indirect responses to grazing. Such a co-existence is theoretically possible even without making assumptions concerning the behaviour of the herbivores, or about their population dynamics (Augner et al. 1991). The evolution of plant chemical defences against herbivores has been analysed by Rhoades (1979).

6.2.2 Induced Defences

Many plants employ induced defences that are activated after attack, rather than constitutive defences which are always present. Such facultative defences may reduce the food quality of a plant, and modify the competitive qualities of grazed and ungrazed species. Current theory predicts that one of the benefits of induced defences is that the plant's outlay on defence, growth and reproduction is allocated in the optimal way. This presupposes that attack is unpredictable and not fatal, that defence is costly and that plants can economise their resources by using them only for growth and reproduction. When herbivores are present, however, activation of the defence increases growth and reproduction. This was tested by examining the effects of damage by the leaf-mining moth *Bucculatrix thurberiella* (Lyonetiidae) on the perennial shrub *Gossypium thurberi* (Malvaceae) of the Sonoran desert. Karban (1993a) found experimentally that the responses induced had a large negative effect on *B. thurberiella*. Miners inhabiting plants that had previously been attacked were less likely to emerge satisfactorily than individuals inhabiting control plants. No evidence was obtained, however, that induced resistance cost the plant anything in terms of survival, growth or seed production (Karban 1993b). This, and other empirical results suggest that the cost-benefit model may not be appropriate to explain the advantages of induced over constitutive resistance.

6.3 Toughness and Spines

The defences of plants against herbivores include the suffusion of the epidermis with lignin, silica, cork or wax so that the tissue cannot easily be bitten or chewed. Other epidermal defences include hairs, glands, spines etc. Some plant surfaces are even endowed with egg mimics which deter ovipositing insects (Crawley 1983).

Although tough and spiny plants are found throughout the world, they are especially characteristic of the flora of desert regions. As in all other defensive mechanisms, none of them renders its possessor completely free from attack. Even the ancient and extremely tough leaves of *Welwitshia mirabilis* (Welwitschiaceae; Fig. 28), endemic to the Namib, may be chewed by Hartmann's mountain zebras (*Eguus zebra*) and black rhinos (*Diceros bicornis*) in time of drought, despite the fact that they also contain distasteful secondary compounds (Sect. 6.2.1). The plant-sucking pyrrhocorid bug *Probergrothius sexpunctatis* is the most common inhabitant of *W. mirabilis*, from which it sucks sap whilst its own haemolymph is sucked by a reduviid bug which lives beside it (Seely 1987). Cacti and *Agave* spp. (Agavaceae) also have tough, fibrous leaves or stems, while the leaves of dragon trees (*Dracaena* spp.) (Dracaenaceae) are also tough and spiky. The leaves and stems of *Fremontodendron* spp. (Sterculiaceae) are festooned with tiny dust-sized burrs which are extremely irritating to delicate human skin and, pre-

Fig. 28. *Welwitschia mirabilis*, showing fibrous leaves (Namib desert)

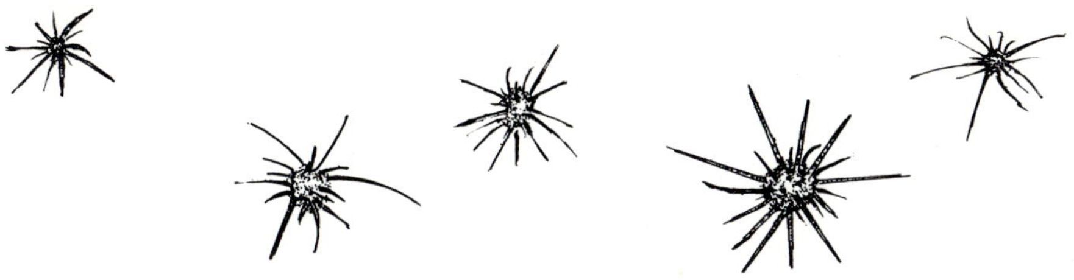

Fig. 29. Stellate hairs of *Fremontodendron californicum* (greatly enlarged)

sumably, to that of other mammals (Fig. 29; J. L. Cloudsley-Thompson, unpubl. observ.).

Spines deter large herbivores, mainly by causing them pain, the mouth, nose and eyes being especially sensitive. Some spines are reinforced by venoms, others, such as the spines of *Zizyphus obtusifolia* (Rhamnaceae; Fig. 30 *left*), damage the tissues of the herbivores by puncturing, ripping or piercing them. Recurved spines, including those of the wait-a-bit thorn (*Acacia mellifera*) of the Great Palaearctic desert, the Namib desert camel thorn (*A. erioloba*), the North American catclaw acacia (*A. greggii*) and ironwood trees (*Olneya tesota*: Leguminosae), impale the flesh of moving animals. Their curvature causes the spines to penetrate deeper as the animals pulls away (Fig. 30 *centre*). Sharp, clean punctures and tears cause less pain than do ragged wounds. Barbed spines, including those of cacti, especially *Opuntia* spp., cause intense pain in humans. These barbs are adapted so that they become firmly attached to the tough, elastic integuments of mammals and birds, and the oral tissues of lizards. Many *Opuntia* spp. have thin membraneous sheaths around their spines (Fig. 30 *right*), the exact function of which is not clear. They may, however, aid in defence by protecting the barbs either from the weather or from becoming clogged with dust particles (Sect. 6.1.1; J. O. Schmidt 1989).

Two aspects of the evolution of typical African plants seem to have been influenced by giraffes, and extinct browsers – the presence of large

Fig. 30. Spines of desert plants. *Left* Piercing spines of *Zizyphus obtusifolia*; *centre* recurved spines of ironwood (*Olneya tesota*); *right* spine of *Opuntia phaecantha* with the sheath removed. Drawn from photographs in J. O. Schmidt 1989)

thorns and spines on many plants, and the characteristic flat-topped shape of savanna *Acacia* spp. Thorns, spines and hooks are also characteristic of *Balanites* and *Scutia* spp., all of which are heavily browsed by giraffes. It seems probable that the spines have evolved as a result of heavy browsing by mammals because in Australia, which lacks large browsing animals, the trees are without thorns (Foster and Dagg 1972; Sect 6.5).

Evidence of the interaction between spines and browsing by herbivores is provided by Young (1987), who found that the thorns of *Acacia drepanolobium* are significantly longer below about 125 cm above ground, where they are browsed by goats, than they are higher up the tree. Controlled experiments have shown that long, straight thorns, in particular, deter browsing mammals. Branches of *Acacia seyal* from which the thorns had been removed suffered greater herbivory from giraffes than did intact branches. Moreover, branches within reach of giraffes produce longer and more dense branches than do higher branches (Milewski et al. 1991).

Spinifex grasses (*Triodia* and *Plectrachne* spp.) are perennial, evergreen, tussock grasses of Australia whose leaves are very harsh and spiny. (The world spinifex is also applied to the grass *Zygochloa paradoxa*, widely distributed in the desert areas of Australia). True spines are stiff, pointed processes, branched or non-branched. Depending upon the organ from which they originate, a distinction can be made between

Fig. 31. *Opuntia* sp., showing spines (Sonoran desert)

shoot spines, stipule spines, leaf spines, root spines and so on. Spines are prominent in desert regions where vegetation is sparse and browsing pressure therefore especially heavy. Some spines, such as those of nettles (Urticaceae) are venomous, but most spiny plants do not belong to groups noted for possessing venomous spines. Nevertheless, the spines of chollas (*Opuntia* spp.; Fig 31) and many of the agaves (*Agave* spp.) cause such intense, persistent pain that it seems quite possible that venom may be involved (J. O. Schmidt 1989).

6.4 Responses to Seed-Eating

Whereas ephemeral plants may, because of their brief period of exposure, not be greatly threatened by herbivorous animals, their numerous seeds are exposed to the attentions of seed-eating animals for lengthy periods. Indeed, whether or not a particular part of a plant – root, leaf, fruit or seed – is eaten is not a matter of chance. In response to being eaten, plants have adapted so that this apparently harmful effect often enhances some aspect of their survival, such as seed dispersal (Sect. 7.5), recycling their nutrients, or prolonging their vegetative life more efficiently than that of their competitors. The nara (*Acanthosicyos horridus*: Curcurbitaceae; Fig. 32) is a spiny shrub with a tangled growth of long sprouts which have thorns but no leaves so that they are almost inedible;

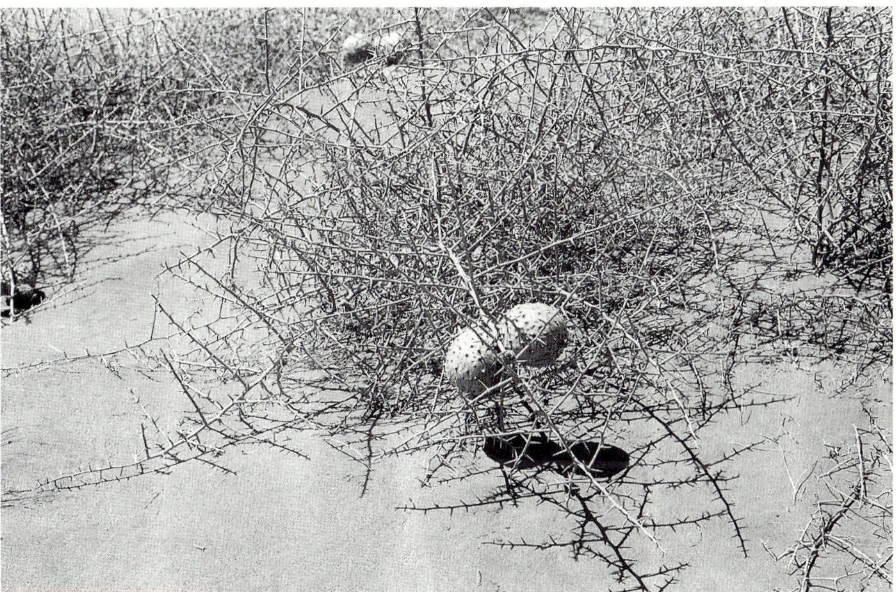

Fig. 32. Nara (*Acanthosicyos horridus*), showing spines and fruit (Namib desert)

at the same time, the fruit, which resembles a melon, is succulent and eaten by a number of animals, as well as by the local Topnar hottentots. In this way the seeds are dispersed (Sect. 7.5). Unlike ephemeral *r*-strategists, species that invest heavily in their seeds or offspring, the K-strategists cannot afford to risk only minimal defences. Seed production is extremely sensitive to herbivore attack, and reduced seed production is a normal reaction to defoliation, see Introduction to Chap. 6.

The relationships between ants and higher plants are often diverse and intricate. Fungus-growing species (unlike fungus-growing termites; Sect. 5.6) depend entirely upon fresh vegetation to provide the substrate for their fungal gardens. They are, therefore, not found in deserts; but the many species that harvest seeds are frequently among the most numerous of the insect faunas of the warmer deserts and semi-deserts of the world (E. O. Wilson 1971). Although the seeds are stored and afterwards consumed, the plants benefit from the dispersal of some of their seeds (Kerley 1991; Gutterman 1993; Sect. 7.5). MacKay (1991) discusses the role of ants and termites in desert communities and their interaction with plants in seed consumption, while seed harvesting, granivory, and foraging by desert ants have been reviewed in depth in another volume of the present Series (Heatwole 1996) and will therefore not be discussed further here.

Seeds of African *Acacia* spp. are usually heavily infested by seed-eating beetles (Bruchidae) and it has been claimed that few of them ever germinate (Crowson 1981); but Miller (1994) found this not to be the case (Sect. 7.5). Indeed, he found that infested and noninfested seeds often germinated equally well. Numerous species of insects avoid plants whose seeds contain enzyme inactivators, but certain Bruchidae breed specifically in such seeds: they have developed enzymes which inactivate enzyme-inactivating substances (Crowson 1981). A study of the distribution of *Acacia* trees and bruchid seed beetles in Saudi Arabia has revealed that different plant-insect pairs occupy a specific range along an altitudinal gradient. At higher levels, seeds of *Acacia negrii* and *A. gerrardii* were infested by *Bruchidius arabicus* while, in the lowlands, *A. ehrenbergiana* seeds were infested by *B. saudicus, A. tortilis* seeds by *B. aurivillii* and *B. sahelicus, A. asak, A. hamulosa, A. oerfota* and *A. seyal* seeds by other *Bruchidius* spp. The more commonly distributed trees supported the largest numbers of beetles, and the percentage of seeds bored into varied with different combinations of species. To complicate matters, the larval Bruchidae were attacked by different species of parasitoids – Hymenoptera: Pteromalidae and Chalcidae (Abdullah and Abulfatih 1995). Apparently, bruchids are not a pest in Australia, which may account for the spread of *A. nilotica* in Queensland (G. E. Wickens, pers. comm.).

When vertebrate frugivores act as pre-dispersal agents, seed-eating beetles adopt a strategy that increases their chances of feeding on the seeds before the fruit has been eaten. A trade-off develops between pre-dispersal and post-dispersal feeding. If the seeds are eaten early, then they tend to have reduced palatability, but late seed-feeders have to find food before dispersal occurs, and there is a risk that the fruits may already have been removed by frugivores. Early dispersal reduces attack by seed-feeding weevils, because any second generation is forced to oviposit on the seeds from which it, itself, has emerged, and there is increased competition. Rapid seed germination helps to prevent attack, but often cannot be achieved in the unpredictable desert climate.

From the viewpoint of the seed-eating beetle larva that would be killed were the seed to be eaten by a mammalian dispersal agent, it is essential to develop early and escape as quickly as possible. If development is too early, however, again the beetle may die, because plants sometimes abort seeds that have larvae inside them. At the same time, if no other seeds are available for the newly emerged adult, it will have to wait in reproductive diapause until a new generation of seeds appears before it can indulge in oviposition: and many trees of the Great American desert fruit only every 5 years or so. Some seeds are scarcely ever eaten. Others, unattractive to beetles, may be susceptible to moths; and the situation is further complicated by the fact that Bruchidae tend to attack seeds that have already been infested by another species. There is more to the dynamics of seed-eating than might first appear to be the case.

The seeds of many Leguminosae contain substantial amounts of the unusual amino acid L-canavanine, which is toxic to most insects, but the distribution of canavanine tolerance among the Bruchidae has not been investigated (Crowson 1981). Other seed-eating animals include weevils (Curculionidae), ants, birds and mammals. So many seeds and seedlings of *Acacia tortilis*, in particular, are eaten that it seems probable that new trees may become established only when populations of rodents and ungulates are reduced by epidemic disease – this also applies to *Faidherbia albida*.

6.5 Thermal Protection

Because of the heat and low relative humidity of the desert, coupled with a low rate of transpiration, plants of arid regions are not often infested by small, sap-sucking insects such as aphids and Aleyrodidae. However, it is believed that the manna which sustained the Israelites during their wandering in Sinai (*Exodus* XVI: 14–15) was the excretion of two species of scale insects (Coccoidea) which feed upon tamarisk bushes. One of these (*Trabutina mannipara*) is found in the mountains; the other (*Najacoccus serpentinus*) is a lowland form (Donkin 1980). Like aphids and other plant-sucking insects that depend upon sap for their nutriment, these scale insects assimilate large quantities of liquid, rich in carbohydrates but extremely poor in nitrogenous content. In order to acquire the minimum amount of nitrogen to balance their metabolism, they must take in an excess of carbohydrates. This they excrete in the form of honeydew, which is harvested by ants. The Sonoran desert cicada (*Diceroprocta apache*: Cicadidae) takes in so much excess liquid that it can maintain a body temperature significantly below the ambient air temperature, using evaporative means (Hadley 1994).

Variations occur between the thermal properties of different species of trees. These influence the numbers and types of arthropods that live on them. Those of the African savanna show adaptations in the structure of their bark. In some, the insulation of the bark is low and the trunks are shaded; in others, the bark is highly insulated and the trunks are not shaded. Trees with white bark avoid overheating of their surfaces by reflecting radiation. The main arthropod groups living on the bark of arid savanna trees are Pseudoscorpiones, Araneae, Collembola, Blattodea, Psocoptera, Coleoptera, Neuroptera, termites, ants and Brachyptera (Diptera). This differs from the faunal composition of the bark of subtropical moist forest trees. Here the proportion of herbivorous and fungiferous to carnivorous arthropod species is about 2:1, whereas in the dry savanna ecosystem the proportion is 1:1. The community structure of the arthropod fauna of the bark on trees is influenced by many factors, including the history and evolution of the ecosystem, as well as by abiotic factors, such as temperature and humidity. The latter affect not only individual species, but also the community structure of the fauna (Nicolai 1989).

6.6 Symbiotic Protective Relationships with Stinging Insects

Myrmecophily, the symbiotic relationship between ants and plants, has long been known. An example is afforded by the whistling thorn (*Acacia drepanolobium*), a familiar tree in dry and arid regions of eastern Africa. The larger brown swellings at the bases of its thorns are inhabited in large numbers by ants (*Crematogaster* spp.), while the swollen spines of the white whistling thorn (*A. seyal* var. *fistula*) are inhabited by other insects. Although the ants do not cause the swellings to appear, they hollow them out by their constant activity. Many myrmecophytes bear special structures for housing ants, and most are furnished with foliar nectaries or other fruit bodies that are exploited by insects (Fig. 33 *left*). Despite the wide range of speciation of mulga trees of the genus *Acacia* in arid Australia, myrmecophytic structures have not been evolved in that continent and spiniform stipules have either not evolved, or been lost (Fig. 33 *right*). In contrast, Afro-Asian species are very spiny and are

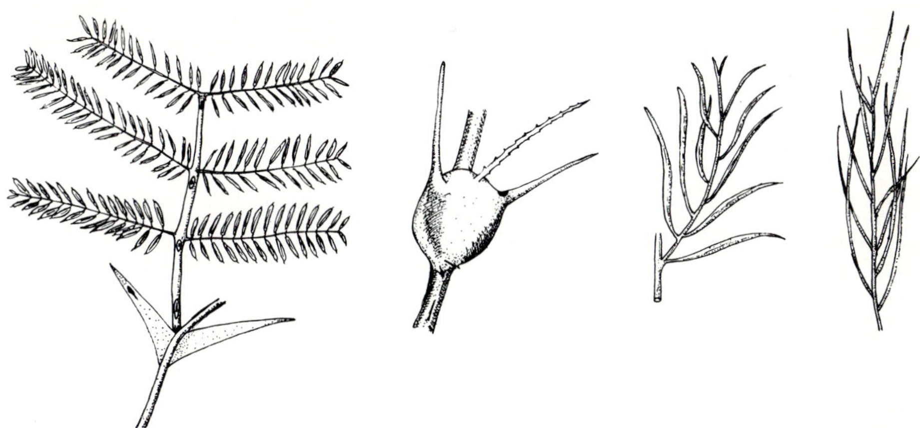

Fig. 33. Leaves and spines of acacias. *Left*: *Acacia cornigera*. Note Celtian bodies on the tips of many leaflets and elongated nectaries on petiole and rachis, which provide for symbiotic ants. The hole near the tip of the spine was made for ants of the *Pseudomyrmex belti* complex: *centre* stipular spines of *A. pseudofistula* with bases swollen into a single domatium for a *Crematogaster* spp. nest chamber; *right* leafy tips of mulga tree (*A. aneura*) showing two different forms of phyllodes and absence of thorns (not to scale). (After W. L. Brown 1960)

strongly myrmecophilic. This is correlated with the presence of large and effective faunas of browsing mammals, both now and in the recent geological past (Fig. 33 *centre*; W. L. Brown 1960).

Foster and Dagg (1972) suggested that the association of thorny *Acacia* spp. with ants is a protective device effective especially against giraffes (Sect. 6.3). Ants tend to concentrate on shoot tips which are preferred by giraffes, whose calves are especially sensitive to *Crematogaster* spp. Moreover, the thorns of *Acacia drepanolobium* are significantly shorter than those of *A. seyal*, a species without ants, suggesting a trade-off between ants and thorns as defences (Madden and Young 1992). In addition to Hymeoptera other than ants, Diptera and Lepidoptera also make use of the large thorns of *A. seyal* var. *fistula* as shelter, while those of *A. elatior* are inhabited by resident populations of ants as well. Lepidoptera form an additional link in the ecological chain. On emergence, the moths make exit holes: ants and other insects then move in, having found ready-made chambers with entrances. Myrmecophily is thus a response to the combined influence of extra-floral nectaries and invasion by caterpillars, both food and shelter being thus ensured (Brahmachary 1982).

6.7 Phenology

Seasonal rhythms are an extremely important factor in the ecology of desert plants, as they are of desert animals (Sect. 3.3). Herbivorous animals need to be attuned to the cycles of their food plants: the same rainfall event that stimulates the germination of ephemeral seeds, and leaf production by perennial phreatophytes, evergreens or succulents, also engenders the appearance and reproduction of the insects and other small animals that feed on them. If there are two rains in the year, each will give rise to a crop of annuals, and each of these will have its own composition: species that belong to the other season remain dormant, despite the moisture. Plants with bulbs and tubers usually produce their stems, leaves and flowers when there is sufficient precipitation. Some do so late in the dry season before any rain falls: the triggering factor is not known.

By having only limited periods of activity, desert plants are able to escape the attentions of non-migratory herbivores that are unable to exist for long periods of time between one meal and the next. When their leaves do appear, all at the same time, they are more than sufficient to satisfy the appetites of larger herbivores that need to feed throughout the year (Sect. 6.2), and happen to be present at the time. There are plenty left over for photosynthesis.

Whether or not a portion of a plant – root, stem, leaf, fruit or seed – is eaten is not a matter of chance. Natural selection has ensured that the harmful effects of being eaten are either minimized, or even put to advantage, as in seed dispersal, recycling nutrients, or competing more efficiently than other plants. The advantage of any particular strategy are only relative and must result in successful competition if they are to persist (R. M. M. Crawford 1989). None of the defensive mechanisms described in this chapter directly improves survival solely in terms of fitness to the physical environment, although some may have dual functions. Spines may reflect radiation in addition to deterring herbivorous animals; stone plants may benefit from both protective resemblance and reduction of water loss. K-strategists cannot afford to risk the minimal defence strategies of highly reproductive grasses and *r*-selected ephemerals. The responses of plants to herbivorous animals are not static: physical and biochemical evolution are continually taking place, and the entire interaction between plants and animals is dynamic. The physiological trade-off between growth and differentiation processes interacts with herbivory and interplant competition to manifest itself as a genetic trade-off between growth and defence in the evolution of plant life history strategies (Herms and Mattson 1992). The co-evolution of producers and primary consumers in deserts is discussed by C. S. Crawford (1981), G. Costa (1995) and others.

7 Community Processes

A community is a grouping of different organisms living together in a particular environment, whose interactions give that community its structure. The advantages of group living are many, and include earlier detection of predators (Sect. 7.1.3), but there may be a disadvantage in arid regions where food and other resources are especially scarce. Groups tend to confuse predators, dilute their effects on the population, and help individual animals from becoming victims (Bertram 1975). Social behaviour in African ungulates has been reviewed by Leuthold (1977) and social organization has been discussed by Delany and Happold (1979). Fluctuations often occur in the sizes of the groups and territories of desert mammals, especially carnivores, which show widespread and extensive flexibility in their social organization (Kruuk and Macdonald 1985; Mills 1990).

In the present chapter, I shall first discuss interactions between individual animals of the same species (see also Sects. 3.9 and 3.10) and then the effects of predators on populations of their prey. Broadening the scope to the interactions between plants and animals, the sequestering of defensive plant metabolites by which animals benefit from plants will then be considered. The benefits to desert plants of pollination and seed dispersal by animals will be debated next, followed in turn by an account of competitive interactions. The chapter will conclude with a review of food webs in the ecobiome. Certain topics, such as parental and reproductive behaviour, and the exploitation of food resources, which have already been discussed at length in the present Series (e.g. Cloudsley-Thompson 1991; G. Gosta 1995; Sømme 1995; Heatwole 1996) will scarcely be mentioned. Community dynamics (Crawley 1983, 1992) is a vast subject that has scarcely been investigated in arid regions and it will not be addressed in the present book.

7.1 Protection of Young and Social Behaviour

Despite some criticism, the basic concepts of *r* and K selection are now generally accepted, although allied desert species can often only be roughly assigned to relative positions along an *r*-K continuum. Although the terms *r* and K selection may themselves be unfortunate, in that they invoke the much-overworked Verhulst-Pearl logistic equation, the concepts that accompany them are independent of that equation and are both clear and useful in the development of new hypotheses in population biology (Pianka 1972). Certainly, as C. S. Crawford (1981) points out, these often-used terms can apply to desert organisms, but other concepts, such as "bet-hedging", may be even more applicable because *r* and K selection are mainly concerned with fecundity and juvenile mortality, which tend to remain constant (Stearns 1976). In deserts, where physical conditions fluctuate greatly, it seems reasonable to expect fluctuating schedules to operate (C. S. Crawford 1981).

7.1.1 Parental Care

Parental care may not only protect eggs and young from climatic extremes (see Riechert 1979), but it also helps to guard them from attack by predators. The subject has been reviewed for desert animals by C. S. Crawford (1981), G. Costa (1995), Maclean (1996) and others. In many instances, it is not possible to separate the basic functions of parental care but, in the case of social insects, no such problem is encountered. The larvae could not survive without it and, moreover, the nest is defended against most aggressors.

When a wasp colony is disturbed, the workers that fly out in defence of the nest operate on an individual level, each one apparently seeking the cause of the disturbance. Moving objects attract most attention, even at a distance from the nest, so it is safer to stand still than to flee from a disturbed colony (Spradbery 1973). Because food supplies are seasonal and widely scattered in the desert, colonial wasps and bees tend to be restricted to oases, where there are reliable sources of nutrition. Termites and ants with subterranean nests are the dominant social insects in arid regions. In their case, defence of the nest is essential for survival of the entire colony. This function devolves to the soldier caste. Thus, the sabre-jawed soldiers of the Saharan *Cataglyphis bombycina* (Formicinae) guard the entrances to their nests with heads facing outwards and mandibles gaping (Délye 1957). Defensive behaviour of both termites and ants has been reviewed by E. O. Wilson (1971), Brian (1983), MacKay (1991) and Heatwole (1996), of termites by S. C. Jones and Nutting (1989), and of ants only by Hölldobler and Wilson (1990).

Social behaviour may be beneficial in some respects but not in others. For instance, guarding the eggs of the field cricket *Gryllus bimaculatus* (Gryllidae) may result in increased infection by protozoan parasites. Simmons (1990) found no correlation between the behaviour of the females and the level of infection of males by gut parasites; but there was a significant correlation between the intensity of guarding by males and the number of parasites that infested them. The males that guarded the eggs most assiduously were those that maintained closest contact with the females; but the latter left heavily parasitised males sooner than mildly or uninfested individuals. This indicates that the females choose in favour of healthy and vigorous males. Selection is therefore pulling in different directions and the advantages of social behaviour are offset by disadvantages of an entirely different kind. Parental care is found in the woodlouse *Hemilepistus reaumuri* and among many subsocial insects and arachnids, including Dermaptera, Embioptera, pentatomid bugs, Hymenoptera, lycosid spiders and scorpions. It has attracted particular attention in the last-named order, in which it has recently been reviewed, among others by Williams (1987) and Polis (1990b).

The costs of reproduction itself are not negligible. Gravid female lizards are more vulnerable to predators than are non-gravid females: not only are their running speeds 20–30% lower (demonstrated in the case of Australian species by Shine 1980), but some gravid females tend to bask more often. Food intake is not reduced, however. Reproductive trade-offs are widespread among reptiles, but the costs vary even among related species. In general, bodily growth is less important in relation to fecundity than the period of survival. This has been demonstrated by experiments on *Leiolopisma entrecasteuxii* (Scincidae), whose main predator is the white-lipped or crowned snake (*Drysdalia coronoides*: Elapidae), and supported by a wide survey of the literature (Shine 1980). Sinervo et al. (1991) found that after parturition, the locomotory performance of female *Sceloporus occidentalis* increased by 20–45% over levels during gestation. In many species of lizards, gravid females experience higher mortality than do other members of the population (Shine 1980). More recently, however, Schwarzkopf and Shine (1992) concluded that the impaired running speed of gravid female water skinks (*Eulamprus tympanum*) may not result in reduced survival because, in this instance, pursuit is a less important determinant of predator efficiency than is detection of the prey. Gravid female skinks are at a greater risk from predators than are males or non-gravid females, but reduced performance is not an important contribution to reproductive cost – at least in terms of predation by snakes and potentially by birds as well. Gravid females allow predators to approach closer than do males or non-gravid females, suggesting that they rely more upon crypsis. Parental care among reptiles has been reviewed by Shine (1988), and trade-off of costs of reproduction in lizards by Schwarzkopf (1994).

The communal nesting system of the ostrich in Kenya has been analysed by Bertram (1992). Not only do both sexes care for their offspring, but some females even allow others to lay eggs in their nests. Of fundamental importance in determining the social system according to Bertram, is the relatively small brain of young ostriches. Perhaps this is what Job (39:17) was referring to when he wrote: "God hath deprived her of her wisdom, neither hath He imparted to her understanding"! It has been known for some while that ostriches breed cooperatively, because the majority of nests contain the eggs of more than one female. Bertram's (1992) key findings are that female ostriches can switch rapidly between reproductive strategies, and that a nesting female can recognise her own eggs. The distinctiveness of the system lies in the ready acceptance by one female of the eggs and young of others. Kin selection plays no part in this, however, and little or no altruism is involved because there is no cost to the major female – she can discriminate among all the eggs in favour of her own. Individual birds adopt different strategies in their own interests, and the pay-off from any one depends upon the strategies that other birds are adopting. Cooperation, competition and manipulation all take place. An important factor is the defence of the eggs and young by the adult birds. Although male ostriches are extremely conspicuous in the open, and can almost be mistaken for buffalo bulls at a distance, the females and young (Fig. 11) are cryptic.

Some desert raptors breed in cooperative groups so that there are more than two adults to defend the nest. According to Wittenberger and Hunt (1985), although social colonies of birds may attract predators, there is a critical density above which numbers suddenly increase too fast to be matched by increases in the number of predators – as in the case of the red-billed dioch (*Quelea quelea*; Sect. 3.9). The main adaptive function of colonial breeding in the quelea is indeed defence against predators (Thiollay 1989). Not only does colonialism provide early warning systems (Sect. 7.1.2), but large colonies swamp local predators with food. They also increase reproductive fitness through social foraging, which maximises feeding efficiency. Although little alarm or defence behaviour, such as communal mobbing (Sect. 7.1.3), is apparent, defence is achieved by breeding in huge, dense and highly synchronised colonies in thorny trees. Most predators, such as kingfishers (*Halcyon* spp.), rollers (*Coracia* spp.) and red-billed hornbills (*Tockus erythrorhynchus*), can swallow only the smallest quelea nestling. Kites, eagles and storks, however, often open nests on the periphery of the colonies, while any young that fall to the ground are eaten by jackals, civets, mongooses, hyaenas, ratels, warthogs and vultures (P. J. Jones 1989). According to Bennun (1994), helpers supply a substantial proportion of the food of nestlings of the grey-capped social weaver (*Pseudonigrita arnaudi*), as they do in the case of many other species of birds that nest communally.

Birds of the family Hirundinidae typically react to predators with a combination of active and passive defence behaviour. Active defence involves diving toward the enemy with repeated alarm calls (Sect. 7.1.2), while passive defense consists of circling above it. Colonial species are less likely to engage in active defence, presumably because they benefit from the presence of large numbers of neighbouring conspecifics, who join together in large flocks that defend their nests communally (C. R.

Brown and Hoogland 1986; Winkler 1994). This correlation probably applies to species that nest in arid regions, such as the mosque swallow (*Hirundo senegalensis*) and the Ethopian swallow (*H. aethiopica*), as well as to species that nest in temperate regions.

The large communal nests of sociable weavers (*Philetairus socius*: Plocidae, Fig. 34) buffer the desert's thermal extremes, enabling the birds to breed throughout the year whenever food is available. These nests may also afford protection from some potential enemies – there is safety in numbers – but this possible advantage is clearly outweighed by the fact that the large nests also attract predators. Cape cobras (*Naja nivea*) can devour all the eggs and chicks of an entire colony of sociable weavers within a few days. The same may apply to nest building by the black-tailed tree rat (*Thalmys paedulcus*), although nothing is yet known about the sociality of these animals, nor about their vulnerability to predators (Lovegrove 1993). The pygmy falcon (*Polihierax semitorquatus*), the smallest of the African falcons, relies exclusively on the huge nests of sociable weavers (Fig. 34) for breeding. These raptors normally live in small family groups of an adult pair and a few juvenile offspring. Their distribution coincides exactly with that of the weavers, and it is unlikely that they would be able to survive without the association with their neighbours (C. Sapsford in Lovegrove 1994). Clearly, pygmy falcons benefit from the association whereas the sociable weavers do not.

Karoo korhaans or black-throated bustards (*Eupodotis vigorsii*: Otididae) are small cryptic bustards endemic to the xerophytic dwarf shrublands of the Karoo. They normally occur in territorial pairs, but up to five individuals may defend a group territory. Groups and their territories are larger in regions of low rainfall. Groups of three or four birds invariably contain a breeding pair and one or two subadult males, which are beneficial in providing assistance with defence of the territory, in which males play the leading role. Male offspring whose maturity is deferred benefit by gaining experience in territorial defence and of territorial life before reaching sexual maturity. In contrast, male bustards, in general, play no part in rearing the young and there is no evidence for co-operative breeding, even though three korhaans have been seen at a single nest (Hockey and Boobyer 1994). Parental care among desert birds has been reviewed by J. L. Brown (1987) and Maclean (1996).

Although relatively common among birds, one of the more highly evolved forms of social organisation in mammals is monogamy (Kleiman 1977), which is found in less than 5% of mammalian species (R. D. Alexander 1974; Wittenberger and Tilson 1980). Although it appears often to be associated with paternal care, there is very little empirical evidence for this from natural populations. Ribble and Salvioni (1990), however, found that, in the monogamous rodent *Peromyscus californicus*, the home range of the male is inversely correlated with population density, which suggests a social influence. Putative fathers

Fig. 34. Sociable weavers (*Philetairus socius*). *Above* Individual birds; *below* communal nest (Kalahari desert)

spend amounts of time in the nest, caring for the young, comparable with the females. This supports laboratory observations on the behaviour of captive individuals.

Unlike other rodents, the spring hare (*Pedetes caffer*) bears only one young at any time, producing three every year; but this is born in an extremely well-developed state. Even then, it remains in its mother's burrow for the first few weeks of its life, benefiting from parental care.

Where they occur in deserts, lions (and possibly alien hyaenas) are the biggest threat to hyaena cubs. Only if several high-ranking female hyaenas are present when a lion attacks is there any resistance. The intruder is then mobbed and threatened while, at the same time, loud whoops rally nearby clan members.

Co-operative behaviour can evolve through kin selection, reciprocity or mutualism. African lions are well known for their co-operative behaviour – they hunt together, rear their cubs communally and defend joint territories. Although the females of a pride are always close relatives, male coalitions are often composed of unrelated animals as well as of close kin (Schaller 1972; Packer et al. 1991). Nevertheless, the males of a group co-operate in response to the roars of rival males – as shown by playback experiments – and this co-operation is conditional neither on kinship nor on the behaviour of companions. Instead, it appears to be based on mutualism (Grinnell et al. 1995). Social organization in Carnivora has been reviewed by Ewer (1973), while Gittleman (1989) argues that for larger species group defence has two functions: protection of the young and defence of valuable food resources.

Eusocial behaviour among insects is paralleled in mammals by the behaviour of the naked mole-rat (*Heterocephalus glaber*; Sect. 3.2). Nearly blind and virtually hairless, mole rats inhabit large subterranean colonies. In these, only one female and her mates, numbering from one to three, conceive all the offspring, while the young from previous litters maintain and defend the groups, as the workers and soldiers in colonies of eusocial insects do (Sherman et al. 1991). In no other mammal, apart from the Damaraland mole rat (*Cryptomys damarensis*), is such a degree of sociality to be found. Eusociality has evidently evolved independently in these two bathyergid genera, which are similar in many ways although *C. damarensis* is not hairless (Jarvis and Bennett 1993; Bennett 1994). There are, of course, no mammals that do not care for their young, but some do so more than others. Social behaviour in fluctuating populations has been reviewed by Cockburn (1988).

7.1.2 Mobbing

Avian mobbing can be defined as an approach towards a potentially dangerous predator, whether it is actively hunting or not, followed by frequent changes of position with most movements centred on the predator. Relatively stereotyped visual and vocal displays often accompany the movements. Mobbing frequently includes swoops or runs at a potential predator, and may even involve direct attack, with physical contact. It is often initiated by a single individual which may then be joined by conspecifics and/or members of other species, the resulting group mobbing the predator simultaneously. Despite the risks involved, mobbers may gain personal benefit from mobbing if the risk of not mobbing is even greater. They may reduce predation on themselves, by informing the predator that it has been discovered, or by chasing the predator away from the mobber's own home range or nest.

In many species, the probability and intensity of mobbing increases during the breeding season and, if it reduces the extent of predation on the mobber's progeny, it must provide direct benefit by increasing the reproductive portion of the personal component of the mobber's inclusive fitness. The occurence of group mobbing outside the breeding season has, however, generated support for the hypothesis that it may sometimes represent a broader form of nepotism or reciprocity. It might provide indirect benefits to a mobber by decreasing the probability of predation on nearby non-descendant kin. Alternatively, direct but delayed personal benefits might be obtained if mobbing were a form of reprocity. Finally, if birds possessed the appropriate population structure, mobbing could be a weakly altruistic trait benefiting the entire group.

According to Shields (1984), the behavioral complex of mobbing in the barn swallow *Hirundo rustica* is a form of parental care (active) as well as self-defence (passive) and mate defence (active). Action defence involves dives toward the predator coupled with repeated alarm calls, while passive defence consists of circling 3–5 m above the predator in a radius of about 15 m whilst issuing alarm calls. The relative importance of these types of behaviour varies between species of Hirundinidae according to their degree of coloniality: colonial species are much less likely to engage in active defence, apparently because they benefit from the pressure of large numbers of neighbouring conspecifics who join into combined passive defence flocks.

An examination of the various costs and benefits of anti-predator behaviour by neighbours in the case of tree swallows (*Tachycineta bicolor*) suggests that the principal cost is the time taken from other activities, especially foraging for the neighbour's own offspring. The principal benefit appears to be the defence of the neighbours' offspring and of themselves, through the distraction and "moving on" of the predators, respectively. Thus, self-interest alone appears to explain participation in group defence, although the lower benefits obtained by the neighbours probably accounts for their lack of active defence (Winkler 1994).

Harassment by mobbing is frequently involved among small birds, especially by the appearance of owls in the daytime. The mobbing calls of birds are characteristically easy to locate, and serve to recruit would-be attackers from far afield. Mobbing is also found amongst mammals, particularly primates. Baboons (*Papio* spp.), for instance, sometimes launch mock attacks against leopards, screaming, charging and retreating. Californian ground squirrels (*Citellus beecheyi*) mob snakes, as do some other species. Mobbing is characterized by conflict between aggression and fear, demonstrated by the alternation between approach and retreat. Displacement activities and other signs of emotion are commonly expressed by mobbers.

7.1.3 Vigilance of Animals in Groups

A negative relationship between the size of a group of socially feeding vertebrates and the vigilance of its individual members is widespread (Sect. 3.10). The main explanation of this is based on the premise that, as the size of a group increases, there are progressively more eyes scanning the environment for predators. Consequently, each individual forager can devote more time to feeding, and spend less on vigilance. The assumption of collective detection is that all members of the group are alerted, as long as the enemy has been detected by at least one individual. For instance, Rasa (1987, 1989) found that in the Taru desert of Kenya, where dwarf mongooses (*Helogale parvula*) are exposed to a high degree of predator pressure, especially from raptors, highly efficient coordinated vigilance has evolved. Subordinate males as a group perform 86.6% of total guarding, and are so effective that 90% of all incipient attacks by raptors are aborted by the guard seeing the predator first and warning the other members of the group. Five adults is the minimum size for a group to maintain its fitness and personal survival. Annual survival rates are significantly greater in packs of eight or more dwarf mongooses than in smaller groups (Rood 1990). The same relationship probably occurs in meerkats as well.

Both solitary individuals and animals in groups need to be vigilant. The question as to which individuals benefit from anti-predatory vigilance in a group of animals feeding together has been addressed by Lima (1994), who points out that it has important implications for the interpretation of the widespread group size effect – the decrease in individual levels of vigilance as the size of the group increases (Elgar 1989). Lima concluded that there may be differences in the probability of escape not only between detectors and non-detectors, but also between vigilant and non-vigilant detectors. It is possible, he suggested, that the latter may spend time in assessing the context in which the group is fleeing, and hence they show a temporal disadvantage in escaping. There are clear potential benefits to maintaining relatively high levels of personal vigilance even in groups (Sect. 3.10).

Animals may sacrifice some of their intake of food if thereby reducing the danger of predation. There is thus a trade-off between foraging and exposure to risk. Stalking predators rely on surprise, and are rarely successful if the prey becomes aware of their presence before the final attack. FitzGibbon (1990) has shown that larger groups of Thomson's gazelles detect approaching cheetahs at greater distances than smaller groups do. Moreover, as mentioned in Section 3.10, where two species differ in their abilities to detect predators, one may gain an extra advantage from joining the other.

Another relationship between vigilance and food intake has been analysed by Illius and FitzGibbon (1994), who concluded that smaller ungulates can better afford to be vigilant than can larger species, which are more likely to suffer time constraints on foraging. The greater need for small animals to be vigilant is accompanied by lower costs. Although this would appear to be self-evident, a note of caution should be introduced. Elgar (1989) claimed that most of the studies that have been carried out worldwide fail adquately to demonstrate an unambigous relationship between vigilant behaviour and group size. There are many interacting variables which make it impossible to conclude with certainty that the amount of scanning behaviour decreases as a result of animals being in larger groups. The most important of these variables are density of the food resource, its distance from a safe place, the composition of the group, the ambient temperature and the time of day. Furthermore, individuals may actually lower their vigilance to the point at which groups do not experience any improved predator detection (Caro and FitzGibbon 1992). Such relationships are, therefore, impossible to assess quantitatively.

7.1.4 Intraspecific Aggression

Animals that live in groups frequently fight over females, territory, food, or merely in establishing their positions in the peck order. Reproductive fighting between males often takes the form of territorial aggression; the owner of a territory mates with the females that enter it and drives rivals away. The sizes of territories may also vary according to supply and demand for food: an eagle needs a larger territory than a hawk. Many mammals mark out their territories, using the scent of urine, excrement, or the scent of glandular secretions. Territoriality is often limited to a definite season and may be maintained without fighting by ritual alone (Eibl-Eibesfeld 1979).

Intraspecific fighting take place among both invertebrates and vertebrates. For example, combat is frequent between males of the Namib desert tenebrionid *Onymacris bicolor*, and the females of rivals are often taken over. Males prefer to mate with larger females, whereas females show no preference for larger males; male size is, however, important when a male is challanged by a rival shortly after copulation. Large challengers are equally likely to challenge other males for large or small females, but small males are more likely to challenge for small females. Large size increases a male's chance of mating, and subsequent males flush out the sperm of their predecessor before leaving their own sperm in its place (Enders 1995). Somewhat similar behaviour occurs in *Physadesmia globosa*, in which species the males not only follow the females but contest with one another for the privilege of mating with them (Marden 1987; Cloudsley-Thompson 1990).

Intraspecific aggression and cannibalism are an important feature of the ecology of scorpions. In the case of equilibrium (K-selected) species such as *Paruroctonus mesaensis*, cannibalism and intraguild predation occur at such high rates that they represent major ecological forces in the evolution of the species (Polis 1980, 1981; 1990b; McCormick and Polis 1990). Cannibalism among Arthropoda has been reviewed by Elgar (1992) and Stevens (1992) and its implications discussed at length.

Some reptiles are aggressively territorial, and injuries such as broken tails, lost toes and various types of scars may result from intraspecific fighting. Many species of lizards have elaborate behavioral signals that indicate the intent to defend their territories. Examples include distinction of the brightly coloured dewlap of *Anolis* spp. (Iguanidae), head-bobbing in Agamidae and so on. Done and Heatwole (1977) noted various degrees of aggression among Australian skinks, of which one of the most aggressive is *Sphenomorphus kosciuskoi*. Its displays consist of flattening and arching the neck and circling the opponent, and fights are vicious. *Ctenotus robustus*, which shows no aggression, lacks the social organisation of *S. kosciuskoi* and *S. quoyi*.

Among lizards, herbivores, in general, are less aggressive than insectivores. The males either defend no territory at all or, if they do, it is small, and subordinates are allowed within it. Females usually confine their aggression to the defence of nesting sites. Herbivorous males have a wider variety of mating systems than insectivorous males. These range from leks (*Agama cristata*: Agamidae) to exploded leks (*Iguana iguana*: Iguanidae), and from large territories with subordinates (*Sauromalus obesus*: Iguanidae) to no territory at all (*Sauromalus varius*). The social behaviour of herbivores in the non-breeding season is more variable than that of insectivores. Herbivores may be highly clumped in dense aggregations (*A. cristata*) or thinly dispersed over wide areas

(*Conolophus suberistitus*: Iguanidae) while clumps of lizards may move through the habitat (*I. iguana*), or have stable distribution (*Ctenosaura pectinata*: Iguanidae and *S. obesus*). It is reasonable to assume in lizards, as in other animals, that a comparatively predictable food source (arthropods) is more conducive to the evolution of territoriality than food resources, such as flowers, fruit and new leaves, that fluctuate in distribution and abundance. The general passivity of herbivores compared with insectivores may well be directly attributable to difference in diets (Stamps 1983). While many species of lizards are territorial, others are hierarachial and some have harems. For all those territorial species studied, crowding results in increased social interaction, increased aggression, and a switch to hierarchial behaviour. The evolution of social behaviour in reptiles has been reviewed by Brattstrom (1974). The widespread occurrence of territoriality in desert lizards is concerned with reproduction and mating, food and foraging, basking and thermoregulation. This subject has been discussed by Heatwole (1976), Stamps (1983), Heatwole and Taylor (1987) and Martins (1994), among others.

Territorial aggression is widespread among birds, and results in spacing and dispersal, whether it be concerned with protecting suitable nesting sites or sources of food. In general, however, it does not appear to be particularly evident in arid regions, where food is scattered and water, when present, is adequate. Among mammals, territorial disputes and aggression are also widespread. The main form of threat by kangaroos (*Macropus* spp.) combines the intention to attack with intimidation, the animals rearing up on their hind feet and, when fighting takes place between rival males, the two animals grasp at each other with their paws, and administer forceful kicks (Sharman and Calaby 1964). In rodents and other small mammals, fighting is relatively unstructured, the two contestants clutching at each other, wrestling and biting. Porcupines threaten their rivals by tail-rattling, while camels (*Camelus dromedarius*) gnash their teeth and salivate so that they literally "foam with rage" (Ewer 1968). Fighting consists of biting and scratching in the Felidae, and the lion's mane may have some protective function. Elephants fight mainly with the trunk, the tusks serving as holding devices. As Ewer (1968) points out, mammals, in general, tend to fight only as a last resort, and death of the contestants occurs relatively seldom.

The most stylised fighting, in which both rivals adhere to a formula, is found in horned species. The horns of giraffes (*Giraffa camelopardalis*) are used during intraspecific territorial fighting. As is to be expected in view of their shape and size, the techniques used are not typical of the fights of other horned mammals. The heads are used to deliver violent blows with a wide swing of the neck while the rivals are facing one another or standing side by side. Bouts of such combat are interspersed with neck fighting in which the two adversaries wrap their necks together and push against each other with their bodies. Such fighting can take place without being fatal because the horns are short and blunt (Innis 1958; Dagg and Foster 1976). The fights of horned and antlered artiodactyls are generally based on a trial of strength by pushing with the head while the horns or antlers are interlocked. When one combatant yields and gives way, he is pursued only briefly and often escapes without damage, although ritualized fighting may escalate and deaths are not uncommon (Huntingford and Turner 1987). Moreover, many variations from the general principle have evolved (Bubenik 1990).

The main function of the horns is not to slash or gore the opponent, but to prevent the heads of the rivals from slipping aside as they push and thrust with all their strength. Branched antlers interlock effectively, but unbranched horns would tend to slip, were there not horny rings on their bases. Some species, such as wildebeest (*Connochaetes* spp.) push in a kneeling position while the horns are crossed rather than interlocked when fighting breaks out among sitatunga (*Tragelaphus spekei*), gemsbok (and other oryx antelopes) and the Indian blackbuck (*Antilope cervicapra*). Among sheep and goats, the initial impact is more important than prolonged pushing. Their fights consist of a series of charges in which the opponents leap at each other, crashing their heads with extreme violence. The horns do not interlock, but have wide bases, and the skull is strengthened to withstand the shock of impact. African buffaloes (*Syncerus caffer*) charge with heads and noses stretched forward parallel with the body. Just before impact, however, they tuck their noses in and take the impact on the bosses of the horns. There are no prolonged pushing matches as in American bison (*Bison bison*). Serious fights are infrequent, but threats are often to be seen (Ewer 1968; Leuthold 1977; Sinclair 1977; Mloszewski 1983; Bubenik 1990).

Fighting among antelopes is usually highly stylised, as we have seen, so that serious harm to the contestants is comparatively rare, but neither waterbuck (*Kobus* spp.), rhebok (*Palea capreolus*), bushbuck (*Tragelaphus scriptus*), kob (*Kobus kob*) nor eland (*Taurotragus* spp.) "keep to the rules": they sometimes injure one another quite seriously and occasionally even fatally. Among artiodactyls, large horns and antlers, used in threat, may be sufficient to deter rivals without the necessity for any violence. Indeed, fighting is usually the last resort among male rivals, and is normally avoided as a result of ritual displays (Spinage 1986).

The shape of the horns reflects the manner in which their owner uses them (Fig. 35). The most dangerous weapons take the form of short, stabbing spikes; but these are dangerous also to the animal that wields them because its neck may be dislocated if it cannot withdraw its horns quickly enough, after driving them into the body of an opponent. Waterbucks lunge at each other with forward-curving horns, sable antelopes (*Hippotragus niger*) hook their opponents with backward-curving horns, while kudus (*Tragelaphus* spp.) wrestle with their horns locked together. Although it is generally considered that the objective of intraspecific

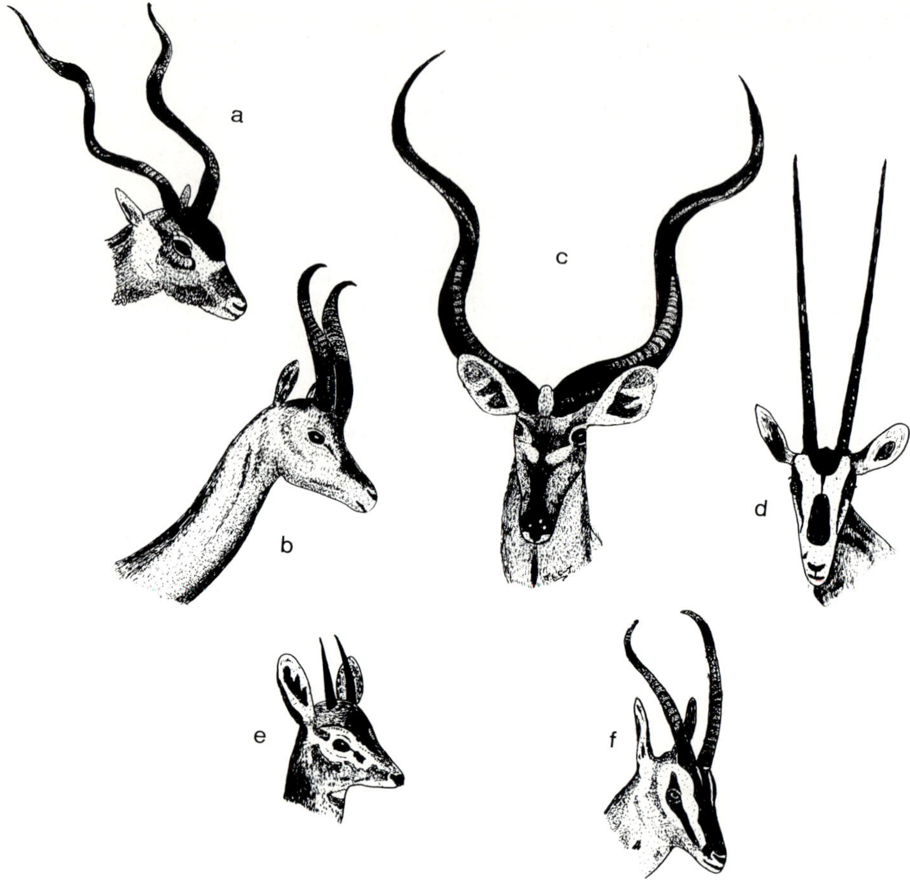

Fig. 35 a–f. Horns of various African antelopes whose shapes are primarily adapted to intraspecific fighting. **a** Addax. **b** Gerenuk. **c** Kudu. **d** Gemsbok. **e** Klipspringer. **f** Soemmering's gazelle (not to scale). (After Gentry 1990 from H. A. Bryden)

fighting among antelopes is to drive an interloper from a territory, there seems to be no inhibition to killing it in the case of waterbuck and the other species mentioned above which follow no set rules. Even among those, however, the victor usually contents itself with a single thrust of its horns. In any case, a seriously wounded animal is likely to be killed before long by a lion, a hyaena, or some other predator (Ewer 1968; Bubenik 1990).

Social dominance among artiodactyls does not always depend upon contests that involve the use of horns. Mule deer (*Odocoileus hemionus*) and other antelope show a kick order, analogous to the peck order of birds, in which fawns and does are subordinate to bucks but are themselves divided into social groups. (Perissodactyla also use their hooves in combat.) In the American bison, a young male must first work his way through all the female ranks and start again at the bottom of the males in order to win himself a position in the male hierarchy (see Cloudsley-Thompson 1965). Emergent principles in the study of agonistic behaviour have been surveyed by Huntingford and Turner (1987).

7.1.5 Interspecific Aggression

Agonistic behaviour is not invariably intraspecific. There are occasions when different species compete with one another for food (Sect. 2.3) as, for example, when lions displace hyaenas or jackals from their kills, or hyaenas chase off vultures and marabou storks (*Leptoptilos crumeniferus*). In addition, herbivorous species may compete amongst themselves for waterholes, or one species may even drive off another for no apparent reason, as when elephants threaten a herd of buffaloes. The latter may hold their ground for a while, but give way eventually to their formidable opponents (Mloszewski 1983). Such interspecific aggression is known as species domination. In a similar way, leopards dominate cheetahs, while elephants, rhinos and buffaloes are higher in hierarchial rank than waterbuck or warthogs (see Cloudsley-Thompson 1965). The adaptive significance of such behaviour is somewhat obscure.

7.2 Effects on Animal Populations of Parasites and Predators

The effects of natural enemies – parasites and predators – on prey or host populations no doubt vary according to circumstances. In some cases, population excesses are removed without long-term effects upon densities; in others, dominant prey species are maintained at densities low enough to prevent over-exploitation of their food resources. Indeed, species richness may even be enhanced as a result of increased predation.

7.2.1 Parasitological Interactions

Not only do parasites harm their hosts directly but, by weakening them, may make them more vulnerable to predators, which often select parasitised rather than healthy prey. Moreover, parasites may influence the nutritional demands of their hosts so that the latter have to spend longer foraging and, therefore, increase their exposure to predators. Foragers sometimes select foods that have anti-parasitic properties. For example, in places where schistosomes are present, baboons often eat the leaves and berries of shrubs that are toxic to schistosomes but, in schistosome-free areas, they do not consume these shrubs (Sih 1993).

The harmful effects of ectoparasites have been discussed in Section 5.2. In many species of mammals these are, to some extent, reduced by grooming – one of the most commonly performed defences. In impala (*Aepyceros melampus*), self-grooming rates are correlated with seasonal fluctuations in the populations of adult ticks. Allogrooming is influenced by the challenge of nymphal ticks, which favour the ear and neck regions which are inaccessible to oral self-grooming (Mooring 1995).

In common with other complex natural systems, it appears that a small number of factors dominate the population behaviour of microparasitic viruses and bacteria, namely their ability to multiply directly and rapidly within the host, the subsequent induction of long-term immunity to reinfection and a mass-action mode of transmission. In the case of macroparasites, there is evidence to suggest that immuno-efficiency is dependent upon the nutritional status of the host. Another surprising interaction between host and prey is seen in the Negev desert parasitoid *Pseudopompilus humboldti* (Pompilidae) and its paralysed host, the eresid spider *Stegodyphus lineatus*. The latter is moved by the wasp from the depths of the nest, where experimental manipulations have demonstrated that neither spiders nor wasp larvae can survive the heat during summer, to the cooler nest entrance – despite increased exposure to visually searching predators, such as birds. This host-storing behaviour is an adaptive trade-off between thermoregulatory requirements and predation risk (D. Ward and Henschel 1992).

Unraveling the evolutionary interactions between parasites and their hosts is extremely difficult because different taxa exhibit varying degrees of specialism and generalism which are even more complicated when both definitive and intermediate hosts are involved (see discussions in Crawley 1992).

7.2.2 Predators and Prey Populations

The dynamic effects of predation upon prey populations generally are discussed by many of the contributors to Crawley's *Natural Enemies* (1992), and although desert animals seldom enter into the discussion, the principles involved have a general application. They will not be discussed in the present book, however. Instead, consideration will be devoted to the interplay between ecological and evolutionary forces in arid environments.

The interactions between harvester ants (*Pogonomyrmex* spp.), which are extremely venomous, and their specialized predators, horned lizards (*Phrynosoma* spp.: Iguanidae), have been analysed in great detail by Schmidt and Schmidt (1989). Their excellent account provides an ideal model for the kind of investigation that will doubtless be carried out in the future on other animal taxa of arid regions.

The huntsman spiders *Leucorchestris arenicola, L. steyni* and *Carparachne aureoflava* of the Namib desert are formidable nocturnal predators (Sect. 2.2), which can capture prey more than twice their own length and width. In addition to moths, beetles and ants, their diet includes at least five species of spiders, three of solifugids, three of scorpions and one each of geckoes and lacertid lizards. Many of these, in their turn, prey on huntsmen spiders, depending on their relative sizes (Hentschel 1994). Such mutual intraguild predation is often found in age/size-structured populations of generalist predators (Polis 1980). Cannibalism occurs among these spiders, as it does among guilds of scorpions (Sect. 7.1.4). As long-lived, territorial, sit-and-wait predators, they may be limited by food supply; but intraspecific interactions must dampen this effect. Very different use of microhabitats by the sympatric *L. arenicola* and *C. aureoflava* may be influenced by intraguild predation, with the smaller spiders occupying areas of soft sand not frequented by larger individuals. The activity of these spiders may therefore be more important for each other than it is for the populations of their insect prey (Henschel 1994).

In the Narra (Nara) valley of the Namib desert, Perrin and Boyer (1994) estimated that rodents remove about 10% of the total invertebrate biomass each month while, on certain interdune plains in the eastern Namib, they consume all the green matter that becomes locally available.

Abramsky et al. (1992) give another example of predator-prey relationships involving rodent-snail interactions. Predation by rodents on snails in the central Negev desert is limited by the availability of shelter for the rodents. When artificial shelters were provided, not only did rodent numbers increase over a period of 5 years, but also their efficiency in the number of snails each one ate. The persistence of the snail-rodent system is due to spatial variation in the relative importance of the direct predator-prey relationship and the effects of immigration by snails. The provision of shelters for rodents increased the heterogeneity of the environment and reduced the risk of predation on the rodents while they were foraging for snails, thereby affecting the behaviour and density of both predator and prey.

The Negev desert snail *Trochoidea seetzeni* (Helicidae) has two coloured morphs, white and brown. Neither of these has any apparent thermoregulatory advantage at different heights above ground either in summer or in winter. Brown snails are more cryptic, but are chosen by their rodent predators more often than expected by chance. The colour polymorphism must therefore be maintained either through some form of external selection rather than predation or temperature, or through a random genetic process, or apostatic selection (Slotow et al. 1993).

The effect of predators on the behaviour of their prey may actually be more important to the ecology of prey communities than the actual mortality inflicted (Kotler and Brown 1988; Kotler et al. 1988; Kotler and Holt 1989). The responses of prey to their predators may even lead to a decrease in the number of prey killed by predators, as increased risk of predation may affect decisions regarding time of activity (Kotler et al. 1988) and the use of microhabitats.

FitzGibbon (1990) noted that of 100 attacks on gazelles by cheetahs, 25 resulted in kills. Males were taken more often than females, but this does not necessarily indicate a preference on the part of cheetahs – males are more likely to be solitary or located at the edge of a group. In addition, males tend to be less vigilant than females. Thus, the behaviour of potential prey affects the diet of predators.

The relationship of the brown and spotted hyaena with other large mammalian carnivores in the Kalahari has been explored by Mills (1990). Because it is almost exclusively a scavenger (Sect. 2.3), the brown hyaena scarcely competes with other carnivorous animals. Only lions and cheetahs have any effect on spotted hyaenas, and this is largely to the hyaenas' advantage, whereas all species, except leopards, affect brown hyaenas one way or another.

7.3 Sequestration of Plant Metabolites

Insects and other arthropods employ a large number of chemical deterrents to predation. In most cases, these are secreted by the animals themselves. Some toxic and unpalatable species do not, however, produce their own chemical defence but, instead, sequester compounds produced by the plants upon which they feed: these are stored in the haemolymph, cuticle or sometimes in specialized organs (Duffey 1980; M. D. Bowers 1990). For instance, the Sodom apple (*Calotropis procera*: Asclepiadaceae) is eaten only by the aposematic grasshopper *Poecilocerus hieroglyphicus* (Acrididae), which sequesters its poisons, and is avoided by other insects, goats and camels (Abushama 1972). For this reason *C. procera* is common around villages in India and Pakistan, in the Sahel savanna, Arabian peninsula and Middle East generally, because every other living plant has been removed by grazing. At the same time, North African desert carnivores seeking a nutrient nibble at the expense of the grasshopper are likely to be greeted by two well-aimed jets of irritating, toxic fluid (Abushama 1984). The congeneric *P. bufonicus* also feeds on milkweed, and Fishelson (1960) found that secretions from its abdominal repellent glands are an effective deterrent against both vertebrates and invertebrates. C. S. Crawford (1981) cites a number of other examples, including the beetle *Timarcha punctella*

(Chrysomelidae), which appears to specialise on *Plantago albicans* (Plantaginaceae) in arid parts of Morocco. Many insects have evolved endogenous chemical defences in addition to chemicals derived from their food because some predators, such as gazelles and hyraxes, have evolved immunity to plant poisons (M. Edmunds 1974).

7.4 Pollination of Plants by Animals

Pollination, the transfer of pollen from the male stamens to the female stigma of a flower, is achieved in a number of different ways. Some plants are pollinated by wind or, rarely, by water. The pollination of flowers is the subject of many books (e.g. Proctor and Yeo 1973; Barth 1985), and will not be discussed further here except insofar as desert species are concerned, since the principles are common to nearly all terrestrial environments. Many species of desert plants are pollinated by wasps, bees, flies, butterflies, moths and other insects whose season of activity, like that of the ephemeral flowers upon whose nectar they depend, is initiated by rainfall: others depend upon birds or bats.

7.4.1 Pollination by Moths

Many of the plants of arid lands are pollinated at night, and depend upon visits by nocturnal moths, of which the different species have characteristic times of emergence. The flowers that are specially adapted to being pollinated by them are mostly white, pale rose, or pale yellow, and are often excellent reflectors of ultraviolet light. These colours show up better than others do in poor light. It has been proved experimentally that the insects known to visit flowers react to contrast either between blossoms and their surroundings, or within the blossoms themselves. In particular, nocturnal moths tend to react most strongly to contrasts between blossoms and their background, whereas day-active species respond more vigorously to multi-coloured flowers.

The blossoms that attract night-flying insects are not only uniformly pale in colour, but the petals are often broken up and strongly dissected so that they contrast markedly with the blackness surrounding them. Some flowers that are pollinated by moths at night are also visited by butterflies during the daytime, thus getting the best of both worlds. Many of them are strongly scented and particularly attractive to hawkmoths (Sphingidae), which fly rapidly from flower to flower and usually take nectar without settling.

There is a tendency for the colours of flowers growing in desert and mediterranean landscapes to shift away from the dominant background coloration so that contrast is enhanced. Pollination among Hymenoptera is highly competitive, and floral colours are well adapted as advertising signals selected by pollinators that learn quickly, while maximum contrast between floral coloration and that of the background is advantageous (Menzel and Shmida 1993).

An extraordinary interrelationship is found between yucca plants (Agavaceae) in arid regions of America, and the moths that pollinate them. All yucca species east of the Rocky Mountains are pollinated by *Tegeticula (= Pronuba) yuccasella* (Incurvariidae). This small, variable species is night-active. The moths spend the day at rest in the flowers of the yucca, which they resemble in colour. Yuccas produce inflorescences of numerous large creamy white flowers, which are scented but smell most strongly at night. Although the yucca moth does not feed, nectar is nevertheless secreted by the yucca flower: it serves to attract insect visitors away from the ovaries of the flowers, and fertilization is effected only by the yucca moth. When the female moth enters a yucca flower, it climbs up one of the stamens and bends its head closely over the top of the anther. This is the part that contains pollen. The moth uncoils its tongue above the top of the anther, scrapes the pollen into a lump with the maxillary palps, and grasps this with its fore-legs. After collecting pollen from three or four stamens, the moth then flies to another flower and inspects closely the conditions of the ovary. The moth is apparently able to tell if the flower is mature, and whether eggs have already been laid in it. If the new flower is suitable, the moth again climbs up the stamens from the base, but this time merely to lay its eggs, boring into the ovary of the flower with its ovipositor. It then climbs the stigmas, which are united to form a tube, and effects pollination by forcing some pollen down this tube. Usually, a single egg is laid in each of three ovules of the ovary and the flower is pollinated after each egg has been laid. Since unpollinated flowers soon die, the behaviour of the moth ensures a supply of food for its larvae. This is provided by the abnormal growth of one or more of the ovules in the neighbourhood of each egg, while the remaining ovules develop into seeds. When fully mature, the caterpillars emerge from the flower and pupate underground (Procter and Yeo 1973). By a series of experimental manipulations C. D. James et al. (1993) showed that only pollination by yucca moths produces mature fruit in *Yucca elata*, and pollination by any other insect would be a rare event.

7.4.2 Pollination by Birds and Small Mammals

Should a snail, in its nightly wanderings, creep across a plant in flower and accidentally transfer pollen from one blossom to another, it cannot be counted as a true pollinator: the chances against successful fertilization in this way would be too great. True pollinators must regularly visit the plants that they pollinate. Insects are the most important agents of pollination in temperate climates, where seasonal changes are pronounced. In the tropics, however, where there is a marked absence of seasonal variations, flowering plants are often pollinated by birds and mammals. This takes place even high up on mountains, where nightly frosts are a regular occurence – showing that it is the evenness of tropical climates rather than their high temperatures which is the factor of greatest importance.

When comparing vertebrates with insect pollinators, two major differences become apparent. First, vertebrates live for much longer and require food all the year round. Many migrate into arid lands at the time of the seasonal rains. Species that feed on nectar are, therefore, restricted to or migrate to regions where there are flowers in blossom. Secondly, birds and mammals require more protein in their diet than insects do. Protein is usually obtained by them from sources other than flowers, for the vertebrates that regularly visit flowers seem nearly always to be attracted by nectar alone and seldom feed much on the pollen. Whereas birds, and especially hummingbirds, honey-creepers, sunbirds and lorikeets, are the most important vertebrate pollinators during the hours of daylight, at dusk their place is taken by night-flying bats and certain other small mammals.

Pollination in the tropics may be effected by small vegetarian mammals. For instance, the baobab or tebaldi tree (*Adansonia digitata*: Bombacaceae) is occasionally pollinated in semi-arid regions of Africa by the bush baby (*Galago crassicaudata*: Lorisidae): the normal pollinators are fruit bats (Jaeger 1954). Species of *Protea* in South Africa are pollinated by nocturnal rodents. The Australian marsupial honey mouse (*Tarsipes spencerae*: Phalangeridae) has become specially adapted for feeding on nectar. Its snout projects forwards and it has lost most of its teeth – those that remain being unusually small, but it has a very long, worm-like tongue with a rough surface which is ideal for collecting nectar from the narrow tubes of flowers (Proctor and Yeo 1973).

7.4.3 Pollination by Bats

The visits of bats to tropical blossoms were observed as early as 1772, but it was only towards the end of the last century that bats were recognised as being significant pollinators. Bat pollination takes place only in the tropics but occurs up to 3400 m above sea level in the Andes. Flower-visiting bats have evolved independently in different tropical regions. In the New World, certain bats (Microchiroptera), whose relatives are insectivorous, have developed fruit-eating habits, while some of them have especially long snouts and tongues for collecting nectar. In *Mansonyc-*

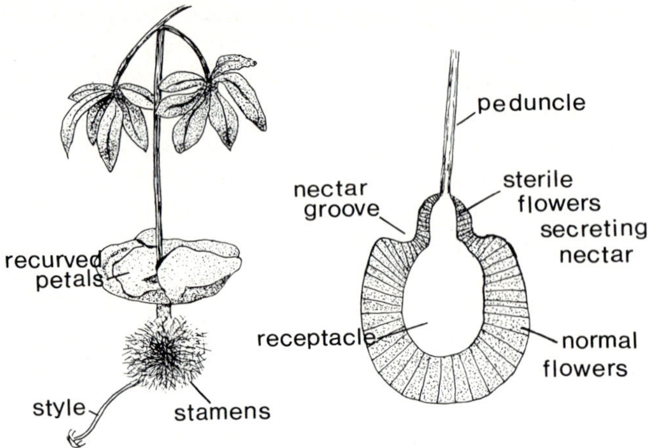

Fig. 36. Bat-pollinated flowers of the New World. *Left* Baobab (*Adansonia digitata*); *right* diagrammatic section through the infloresecene of *Parkia clappertoniana* (not to scale). (After Proctor and Yeo 1973; after Jaeger 1954; Baker and Harris 1957)

teris harrisonii, for example, the tongue is almost as long as the remainder of the body – some 8 cm in length. At the same time, the hairs of the body have scales which enable them to transport pollen from one flower to another, just like the scales on the hairs of bees. Some of the fruit-eating bats or flying foxes (Megachiroptera) of the Old World have likewise become adapted to feeding on flowers. The snout is long, the teeth reduced, and the tongue extensile and elongated. These bats are presumably colour-blind because the retinas of their eyes have many rods but no cones. The sonar apparatus, by which insect-eating bats detect their prey, is not well developed, but the sense of smell is unusually acute. Both nectar and pollen are eaten by these animals.

The chief characteristics of bat-pollinated flowers are that they open from dusk to dawn; they usually have a disgreeable, sour and musty scent that attracts the bats; the colour is often dingy, but they are strong enough to bear the weight of a bat; they produce a great deal of sticky nectar and pollen; and their position is unusually exposed on bare branches, trunks and long dangling pedicels or stalks. There are two main kinds of bat-pollinated flower. One is bell-shaped, the other has a gullet-like form. The bell-shaped and gullet-like flowers are both visited for their nectar, but the latter are asymmetrical in shape. A good example is provided by the African sausage tree (*Kigelia africana*: Bignonaceae). The flowers are dark in colour, fleshy, sour-smelling, and open only at night. They hang down from long stalks and the lower lip is wrinkled: this affords a grip for alighting bats, which crawl inside to get nectar and thus pollinate the blossoms. The bell-shaped flowers of the baobab tree also hang down with their large white petals bent backwards. The numerous stamens form a tube below the flower. They end in a tuft covered with purple anthers bearing pollen on which bats alight to lap up nectar from the base of the petals (Fig. 36 *left*).

Parkia clappertoni (Leguminosae) is another African savanna tree whose pollination by bats has been studied in detail (Baker and Harris 1957). Its inflorescence consists of a mass of tiny flowers, densely packed together to form a solid ball (Fig. 36 *right*). The flowers are red at first, then they become purple and, finally, salmon pink. They produce a weak, fruit-like scent and lack the unpleasant odour of *A. digitata, K. africana* or the West African kapok tree (*Ceiba pentandra*: Bombacaceae). The staminate inflorescences of *P. clappertoni* are visited at dusk and dawn by vast numbers of honeybees (*Apis mellifera*) and also at dusk by the fruit bat *Epomophorus gambianus*: *Nanonycteris veldkompii* takes over later in the night. Fruit bats cling to the flowers while drinking the nectar and afterwards, inadvertently, transfer pollen to other inflorescences. Nectar constitutes only a small proportion of their diet.

The possible origins of bat pollination have interested a number of naturalists. In the case of the New World fruit-eating Microchiroptera, it seems possible that flowers were visited in the first instance in order to collect the insects on them. Later, some insectivorous bats took to eating fruit or nectar but insects and the juices of fruits are sometimes found together in their stomachs. The development of pollen-retaining scales on the hairs of these bats has not been explained. Nectar-feeding, and consequent cross-pollination of flowers, must have evolved somewhat earlier among Old World fruit-eating bats because these are more highly specialised – with pointed snouts, long tongues and reduced dentition.

Many Old and New World bat-pollinated flowers have similar smells. At first, this may seem rather surprising because the relationship between bats and flowers has evolved independently in the two regions. It has been suggested, however, that the scent of bat-pollinated flowers may have been evolved in imitation of the acrid smell of the glandular secretion which most bats produce, and by which they find one another when they gather in flocks. This would explain the fact that the flowers of the African sausage tree, normally pollinated by Megachiroptera, are also attractive to New World nectar-feeding Microchiroptera (Procter and Yeo 1973).

7.4.4 Floral Reproductive Mimicry

Few plants, if any, apart from insectivorous species (which do not normally occur in arid regions), appear to have evolved mimicry in response to nutritional problems. Reproduction mimicry, however, occurs in species that utilise false sensory cues which attract animals that effect pollen transfer. The greatest development of mimicry in plants has evolved in consequence of its extensive presence among Orchidaceae (Wiens 1982).

Whilst batesian and mullerian mimicry among animals have been studied extensively (Sect. 3.6), floral mimicry is little known. Indeed, the term batesian mimicry has been rejected for plant pollination systems on the grounds that floral mimicry involves attraction, rather than repulsion of the signal receiver. Nevertheless, both batesian mimicry in palatable animals and floral mimicry in plants with non-rewarding flowers are the outcome of a common evolutionary process. Selection favours resemblance to an unpalatable animal, or a rewarding plant model, respectively, because the fitness of the mimic is increased when it is perceived by a predator or pollinator, respectively, to be an example of the model. For this reason, the category of batesian mimicry should include floral examples (Johnson 1994).

Much of the evidence for batesian mimicry in plants consists of anecdotal information about floral symmetry and the sharing of pollinators (Little 1983; Dafni 1986) and only a few studies contain experimental evidence based on the predictions that: (1) the mimic and model should occur at the same place and time except when signal receivers have long memories; (2) the mimic should occur at a lower frequency than the model; (3) the mimic should resemble the model sufficiently closely for the signal receiver to be unable to discriminate between them and (4) the fitness of the mimic should be higher in the presence of the model than in its absence. One instance that fits all these requirements is the batesian reproductive mimicry of *Disa feruginea* (Orchidaceae) by *Tritoniopsis triticea* (Iridaceae) in South Africa. Johnson (1994) has shown that *D. ferruginea* depends entirely upon the butterfly *Meneris tulbaghia* (Nymphalidae) for pollination. Although its flowers contain no food reward, they secure pollinator visits on account of their resemblance to the red, nectar-producing flowers of *T. triticea*; while an orange-flowered form mimics the orange nectar-producing flowers of *Kniphofia uvaria* (Asphodelaceae). Analysis of the foraging movements of the butterfly in a mixed stand of *D. ferruginea* and *T. triticea* proved that it does not distinguish between the two plants. Moreover, populations of *D.ferruginea* which are sympatric with *T. triticea* have higher levels of pollination and fruit production than have populations where the orchid grows alone.

Unidirectional exploitation of one plant by another can be explained as either that the model donates energy needed for pollination by the mimic or that the pollinator is not rewarded when visiting deceptive flowers and the adaptation of the mimic acts on naive pollinators. Olfactory cues are decisive when optical stimuli fail, and vice versa (Dafni 1986).

The flowers of *Huernia zebrina* (Asclepidaceae), for instance, lying on the desert sands in southern Africa, smell intensely of carrion. They are often larger than the whole of the rest of the plant, and strikingly beautiful. The pronounced concave ring in the centre of the blossom resembles, in colour, a bar of chocolate. No doubt the combination of smell and colour serves to attract the attentions of pollinating flies, as do the brimstone-coloured flowers of *H. hysterix* with their red spots and stripes (Höln and Petermann 1980). Many parallels are to be found in mesic regions.

Mimicry and the deception of pollinators have been reviewed by Dafni (1984), but this author does not refer to anything peculiar to the plants of arid regions.

7.5 Seed Dispersal by Animals

The seeds of a number of desert plants, such as *Reboudia pinnata* (Cruciferae), *Trigonella stellata* and *Astragulus tribuloides* (Leguminosae), *Aizoon hispanicum* (Aizoaceae) and *Plantago coronopus* (Plantaginaceae) are dispersed by wind; while those of others, including *Blepharis ciliaris* (Acanthaceae; Fig. 26) and *Anastatica hierochuntica* (Cruciferae), are dispersed by rain. Their mechanisms of release have been described in detail by Gutterman (1993). Porcupine (*Hystrix indica*) diggins and other depressions in the deserts of Israel form a wind trap for the seeds that are blown across the bare soil surface (Gutterman 1982).

An even more important aspect of the biotic interrelationships between plants and animals is zoochory, or the dispersal of seeds by animals. In some years, more than 70% of seed production is gathered by granivorous mammals, birds and ants in certain desert areas of Australia, North and South America and the Great Palaearctic deserts. Both rodents and ants remove seeds in large numbers in many arid regions (Sect. 6.4; Gutterman 1983, 1993). Rodents have been found to be even more efficient than ants (J. H. Brown et al. 1975). Although large numbers of seeds are eaten, many escape, and later germinate to establish new plants some distance away from where they were originally collected.

The pods of *Retama raetam* (Fabaceae), a large evergreen shrub of North Africa and the Middle East, are eaten by hares (*Lepus capensis* and *L. europeus*) and goats. The undamaged seeds are later to be found in their droppings, and germinate much faster than do seeds that have not been eaten, few of which germinate at all. Similarly, seeds of *Acacia* spp. (Mimosaceae) pass through the digestive systems of gazelles, ibex (*Capra ibex*), Bedouin goats and camels. Again, germination is much higher after the seeds have been eaten. Seeds voided rapidly are far more likely to germinate than are those that remain for longer in the guts of animals (Murray et al. 1994). This explains the presence of laxatives in the fruits and pods of plants such as *Cassia* spp. *C. italica* is reputed for its medicinal value, and yields the official *folia* and *fructus sennae* used as a purgative and stimulant (Batanouny 1981).

The suggestion has been made that the function of the purgative is to prevent the seeds from being eaten by herbivores. It would be surprising if a goat or gazelle, after several hours of browsing, should be able to know which particular plant had later caused it discomfort. For this reason, it seems that the function of the purgative is not deterrence of herbivores (Sect. 6.2.1) but engendering rapid passage through the alimentary canal of the animal that ate the seeds. The fleshy fruits of *Carpobrotus* spp. (Aizoaceae), known as suurvye or sour "figs", are readily eaten by mammals in the karoo of South Africa. They are remarkably sticky, often adhering to the feet of birds and being transported considerable distances from the parent plants. So, too, are the sticky fruits of vaalsandbietou (*Chrysanthemoides incana*: Compositae) and *Eriocephalus* spp. (Compositae), the common perennial shrubs known as kapokbossie (Lovegrove 1993).

Seeds and the fruits containing them are also eaten by many species of birds. For example, in Israel and Sinai, the small black berries of *Lycium schweinfurthii* (Solenaceae) are dispersed by birds (Danin 1996). In arid Australia, the ripe fruits of the nitre bush (*Nitraria billardieri*: Zygophyllaceae) are consumed in great numbers by emus (*Dromaius novaehollandiae*: Dromaiidae). The seeds are highly resistant to digestion, while percentage germination increases considerably after they have been eaten: 67% emu-ingested seed germinated after 24 days, compared with only 17% of hand-collected seed. Germination of ingested seed was also much quicker (50 vs. 3% in 4 days). Emus apparently facilitate germination by removal of the salt-rich endocarp, but mammals are far less effective in this. According to Noble (1975), who made these observations, emus apparently play an integral role in the successful establishment of nitre bushes where heavy clay soils are dominant and seed is rarely buried. On light-textured soils, however, seed germinates readily without being ingested – provided that it has first been covered by windblown sand.

Seeds of *Acacia tortilis, A. nilotica* and *A. hebeclada* germinate equally well whether or not they have been infested by Bruchidae (Sect. 6.4), but ingestion of the pods by large herbivores such as giraffe, kudu, impala (*Aepyceros melampus*), steenbok (*Raphicerus campestris*), grey duiker (*Sylvicapra grimmia*) and ostrich lowers bruchid infestation compared with uningested seeds. Uninfested, ingested and voided seeds germinate significantly better than do uninfested, uningested seeds. Furthermore, infested *A. tortilis* seeds germinate better after being defaecated than do infested but uningested seeds. Ingestion of seed pods by large herbivores may therefore reduce bruchid infestation, increase germination, and consequent seedling recruitment of *Acacia* seedlings (Miller 1994).

The seeds of *Iris petrana* are shaded with arils that are attractive to ants, while the fruits of mistletoes are covered by a sticky substance which adheres to the beaks and legs of birds so that the seeds are carried from one tree to another and germinate on the branches without requiring water (see Gutterman 1993).

The seeds of many desert grasses (Graminae), such as *Cenchrus* spp., have spikelets that adhere to the fur and hair of passing mammals, while in *Aristida* spp. the callus at the base of the spikelet readily attaches itself to the fleece of sheep and the hair of goats and gazelles. Fruits in which the pericarp is provided with adhesive spikes or hooks are more abundant than those with modification to the bracts, branches, calyx or styles. *Clypeola* spp. (Cruciferae) of Central Asia possess hairs that have evolved into spines which adhere to mammals, as do many other Cruciferae of desert regions, such as *Euclidium* spp. The woody pods of *Proboscidea* spp. (Pedaliaceae), the "mule grab" of Texas, and "devil's claw" of the Kalahari, are about 7.5 cm long and fusiform. They end in two long, sharp claws each about 15 cm in length and hooked at the tips. By these they are held to the hooves of the ungulates that transport them. (For an old, but comprehensive account of seed dispersal, see Ridley 1930.)

Hooked, barbed or viscid adhesive fruits are especially characteristic of woodlands, disturbed habitats and dry regions, according to Sorensen (1986), who reviewed the subject of seed dispersal by adhesion. Large seeds require special protective devices - which are attractive to herbivores - and also special dispersal mechanisms. Hooked and barbed fruits are therefore selected because they are both protected and adapted for dispersal by birds and mammals. In true desert, however, there is little advantage in being scattered; and only 1% of 604 species in the desert regions of the world are dispersed by adhesion to animals. The openness of desert vegetation and extreme climatic variation favour antilechory (mechanisms that hinder dispersal), but antitelechory is a side effect whose adaptive value is not directly related to dispersal. There is a higher turnover of microscale spatial patterns among antitelechoric species which - especially basicarps - are widespread and dominant in the desert regions of Israel (Ellner and Schmida 1981; see discussion in Gutterman 1993; seed dispersal among desert plants has been reviewed by Mott 1979).

7.6 Cooperation Between Plants

Certain desert plant species, often the larger co-dominants, are uniformly spaced, while smaller species may be clumped in their interstices. Sometimes there are associations between adults of one species and juveniles of another, giving rise to the concept of "nurse plants". An example is afforded by the cholla cacti (*Opuntia* spp.) of the Mojave desert in southern California. The youngest chollas are strongly associated with particular nurse-plant species, especially the perennial grass *Hilaria rigida*, but differentially among cholla species and between sites with different compositions of potential nurses. Nurse plants are usually replaced by growing chollas, but some recur and are associated with cholla nurses (Cody 1993).

Establishment of the saguaro cactus in the Sonoran desert is largely limited to areas beneath the canopies of nurse trees and shrubs. The close proximity of saguaro cacti, however, leads to a relative increase in stem die-back, as well as to greater mortality in a common nurse tree, the foothill paloverde [*Parkinsonia* (= *Cercidium*) *microphylla*: Leguminosae], indicating that the nurse plant is actually harmed by the association. The establishment of perennial plants beneath the canopy of other plants is one of the principal components of the dynamics of desert plant communities (McAuliffe 1984). Not only does this apply to saguaro cacti in the Sonoran desert but, in the Namib, a leaf succulent, *Aloe asperifolia* (Aloaceae), and the stem succulent *Euphorbia damarana* (Euphorbiaceae) are locally common in the transition zone between the arid grasslands of the eastern desert and the hyper-arid peneplains of the coastal zone. *E. damaran* reproduces by seeds which apparently require sheltered sites for germination, and these are provided by the aloe rosettes. The aloes may gain marginal benefits, such as shade and protection from predators, but produce fewer flowers and rosettes when in association with euphorbias (Dean et al. 1992).

Pantastico-Caldas and Venable (1993) found equivalent competitive effects between the Sonoran desert annuals *Plantago patagonica* (Plantaginaceae) and *Pectocarya recurvata* (Boraginacea) in three different habitats in a creosote flat, viz: sandy rivulets of tertiary wash and arroyo, sandy clay soil at the base of a small hill, and south east of a gently sloping (9°) hill. Target plants were scored for survival, fecundity and size, and it was concluded that the habitat-density interactions which differ among different species may actually promote their co-existence.

This point was demonstrated in a study by Silvertown and Wilson (1994) of the community structure of shrubs in the Chihuahuan desert of Mexico. Creosote brushes (*Larrea tridentata*: Zygophyllaceae) were the shrubs most frequently found on small patches, suggesting that their presence was required before other species, such as *Opuntia rustreva*,

O. schlottii/grahamii and *Jatrophoa dioca*, became established. All species were found in species-rich patches, in the colonization of which facilitation, inhibition and tolerance all play a role. Barbour (1981) has reviewed earlier work on positive associations among desert plants, while Danin (1996) discusses mycorrhizal relationships among the plants of desert dunes.

7.7 Competitive Interactions

Competition is probably the most important factor in community relations: intraspecific competition is logically the keenest density-dependent limitation to population increase within a species. All living organisms compete for various limiting factors – space, territory, nutrients, water, members of the opposite sex and so on. They compete with members of their own species, with members of other species and, both indirectly and even directly, there may be competition between plants and animals. Competition inevitably results in natural selection, and permeates every aspect of desert ecology.

Most documented cases of competition are actually not entirely reciprocal, because one of the species has a markedly greater effect on the fitness of the other (Crawley 1983). The temporal displacement of seed-harvesting ants in the Australian deserts, for instance, may be a response to aggression from non-harvesting ants of the genus *Iridiomyrmex* (Morton and Davidson 1988). Competition between ants has been discussed in some detail by Heatwole (1996) and will therefore not be considered further here. It seems probable that competition may be more significant in some taxa than it is in others. The main types of competition are interference competition, exploitative competition and diffuse competition (Heatwole and Taylor 1987), but they have not been distinguished or discussed separately in the present chapter.

7.7.1 Competition Amongst Plants

The competition among plants for water, space, light and mineral nutrients is obvious under many conditions. The wide range in carbon isotope discrimination values (Δ) maintained between neighbouring shrubs of *Encelia farinosa* (Compositae) suggests a trade-off between conditions favouring survival in both periods of long-term drought (low Δ) and wet years (high Δ) (Ehleringer 1993). There is also competition for pollination, if pollinators are scarce, and for effective mutualism if one of the mutualistic pair lacks abundance locally. Competition for space is reduced when water, rather than light, is the limiting factor.

Competition between desert plants sometimes even extends to the synthesis of allelopathic chemicals that inhibit the growth of other plants. For example, the leaves of *E. farinosa* in the Colorado desert, secrete 3 acetyl-6 methoseybenzaldehyde, a plant growth inhibitor (R. Gray and Bonner 1948a, b); while the guayule (*Parthenium argentatum*: Compositae) secretes from its leaves a substance (trans-cinnamic acid) which is toxic to seedlings of its own species (Bonner and Galstonn 1994; autotoxicity has been reviewed by Rice 1984). *Thamnosoma montana* (Rustaceae), *Sarcobatus vermiculatus* (Chenopodiaceae), *Prosopis juliflora* (Leguminosae), as well as *L. tridentata* and *Viguiera reticulata* (Compositate), natives of the same desert region, all secrete compounds that are toxic to other plants (Bennet and Bonner 1953). Again, the pericarp of mesquite is autotoxic to seed germination and seedling growth due to the presence of allelochemicals (Warrag 1994). The lignan nordihydroguaiaretic acid present in the leaves of *Larrea tridentata* exerts an allelopathic effect on the surrounding vegetation. Long-chain fatty acids formed by *Polygonum aviculare* (Polygonaceae) suppress Bermuda grass (*Cynodon dactylon*: Graminae), and terpenoids in *Eucalyptus globulus, E. camaldulensis* (Myrtaceae) and *Artemisia absinthium* (Compositae) are also allelopathic agents. In the Californian chaparral, the bare zones surrounding individual shrubs or clumps of *A. californica* and *Salvia leucophylla* (Labiatae) are due to the allelopathic effect of volatile terpines (Harborne 1993).

Some plants do not require the use of toxins in order to compete successfully with, or prevent the growth of, other plants. Two common perennial shrubs dominate the landscape in the Karoo of southern Africa. These are *Eriocephalus ericoides* and *Pentzia incana* (Compositae), and they owe their success to the wasteful use of groundwater whenever rain falls. The stomata of their leaves are then fully opened, and much of the water that has been taken up is lost to the air by transpiration. According to G. Midgley (in Lovegrove 1993), by denying nearby plants and seedlings access to water, established shrubs prevent them from growing – or their seeds from germinating. Furthermore, the rapid turnover of water ensures the assimilation of most of the nutrients present in the soil. Mesquite (*Prosopis* spp.) bushes likewise compete for moisture. They are widely spaced, growing outwards to form rings as the centres die. Large rings, more than 25 m in diameter, may be many thousands of years old.

Nearest-neighbour analysis indicates that strong competitive interactions occur between the perennial grasses growing on the sides of sand dunes in the Namib desert. These operate both intraspecifically within a vegetation zone, and interspecifically where one species replaces another over the dune slopes. *Stipagrostis sabulicola* apparently stabilises the shifting sand of the upper dune slope through a process of mound-building, and is replaced over time by *S. namaquensis*. The process is evidently a cyclical succession, since strong winds bury the mid-dune species. *S. namaquensis* and *Cladoraphia spinosa*, causing their eventual death (Yeaton 1990).

The majority of woody plants in the Mojave desert are small woody shrubs representing numerous families. Observed neighbour frequencies show that some species are "preferred" neighbours, while others are avoided as neighbours. It has been suggested that the likely mechanism of the spacing patterns observed is differential compatability of root systems, which may be important in the maintenance of diversity in desert shrubs (Cody 1986). In a study of the effect of neighbouring plants on root distribution in *L. tridentata*, Brisson and Reynolds (1994) found that the major development of root systems is away from competitive pressure from neighbouring bushes. There is a relationship between the shapes of the root systems and the integrating effect of competition. This results in a pattern that reduces overlap between the roots of neighbouring plants. Competition for space occurs only when a root system is crowded on all sides. A plant may show plasticity in the response of its roots to interaction with those of its neighbours but, by doing so, it may have to pay a physiological price that reduces its overall performance.

Autotoxicity, expressed as the inhibitory effects of the pericarp on seed germination and seedling growth of mesquite, was investigated by Warrag (1994). Although the inhibitory effect of any plant extract on the same or another plant species could be attributed to its low pH, low osmotic potential and/or naturally occurring phytotoxins or allelochemicals, it was shown unequivocally in this case to be due exclusively to the presence of water-soluble phytotoxins.

Finally, desert plants may even compete with one another indirectly by supporting the herbivores or pathogens of the other species (Fowler 1986). The composite shrub *Gutierrezia serothrae* excludes the composite forb *Machaeranthera canescens* from some sites by harbouring a grasshopper which eats both species, but to different extents (M. A. Parker and Root 1981). Another complex interrelationship between plants and the herbivores that feed on them was illustrated some years ago in the control of prickly pear (*Opuntia* spp.) in South Africa. This was hindered by baboons feeding on cladodes of the cactus infested by the moth (*Cactoblastis cactorum*: Pyralidae). Plant-plant interactions have been reviewed by Barbour (1981).

7.7.2 Intraspecific Competition in Animals

In the present section, competition that does not invoke aggression will be discussed. A few examples only will be cited because the subject has already been discussed in other volumes of this Series (G. Costa 1995; Heatwole 1996).

Populations of *Helix texta* (Helicidae) in Israel usually consist of a number of mature adult snails and numerous immature juveniles. Very few snails of intermediate age are present. Only when some catastrophic event occurs, such as predation by wild boars, is the number of adults significantly reduced. In these circumstances, the juveniles develop at an accelerated rate. Although a high survival rate is generally found among immature individuals, their development is suppressed in balanced populations. The existing adults must therefore exert some form of inhibition on the development of other snails. A growth-inhibiting factor probably operates via the mucus secreted during locomotion. This mucus may contain an inhibitory pheromone which prevents overcrowding (Heller and Ittiel 1990).

Tadpoles of the spadefoot toad (*Scaphiopus couchii*) respond adaptively to variation in food resources. Where these are greater, they metamorphose at a larger size than in ponds where their density is higher. Where there is less food, however, they metamorphose at a uniformly small size but vary in the time required to reach it (Newman 1994).

Although much is known about how animals behave once their territories have been established, relatively little is known as to how those territories are acquired in the first instance. In an experiment by Stamps and Krishnan (1995), arboreal, vegetarian lizards (juvenile *Anolis aeneus*: Iguanidae) were released into patches of favourable microhabitat in the field: their social interactions and use of space were then recorded throughout the period of settlement. The results obtained supported the hypothesis that territory is won, not by contests of dominance, but by persistence. Not only did the juvenile lizards that were released fail to acquire the areas they won by agonistic interactions with their opponents, but both winners and losers subsequently avoided the locations of their encounters. Dominating individuals gained territories by repeatedly chasing subordinates away, while the latter captured space by persisting in the face of repeated attacks by dominants. The amount of space taken by a subordinate was inversely related to the rate at which that subordinate was attacked by the dominating *A. aeneus*. Success in the acquisition of territory by persistence has been demonstrated in a number of other animal taxa, not necessarily the inhabitants of arid regions (Stamps and Krishnan 1994a, b). Establishment of territory by intraspecific aggression has already been discussed above (Sect. 7.1.4). As in some raptors, where dimorphism prevents intrasexual competition for food, honey badgers (*Mellivora capensis*) show striking dichotomy in the sizes of food items taken. They either dig up scorpions, lizards, and small rodents, or they prey on aard-wolves and bat-eared foxes (*Otocyon megalotis*; Kruuk and Mills 1983). Zemel and Lubin (1995) have calculated that when resources are scarce and patchy, intergroup competition tends to keep animals in their groups, even when these groups are overpopulated. Consequently, animals tend to forage in groups that are larger than the expected optimal size.

7.7.3 Interspecific Competition in Animals

Behavioral interactions between sympatric species, other than those associated with predation, may occur by chance; for example, when two or more species aggregate at waterholes, or feed in the same grazing area. In contrast, positive interactions result when associations are mutually beneficial, as in communal vigilance (Sects. 3.10 and 7.1.3). Positive associations also obtain when oxpeckers (*Buphagus* spp.: Sturnidae) perch on the backs of large mammals and feed on their ectoparasites; when vultures feed on quelea nestlings swept from their nests by eagles (Sect. 7.1.1), when warthogs (*Phacochaerus aethiopicus*: Suidae) feed on *Balanites* fruits shaken from the trees by elephants (*Loxondonta africana*); or when elephants dig holes in dry river beds, which provide water both for themselves and for other animals (Delany and Happold 1979). Many examples are known of commensal behaviour among mammalian herbivore species. For instance, in East Africa, zebras graze first and remove taller stems, wildebeest then crop the leafy grasses, while gazelles nibble the short sward of herbs: but the numbers of one taxon are not dependent upon those of the others (Crawley 1983).

Even in the comparatively simple desert ecosystem, intraspecific interactions can be exceedingly complex. For instance, scorpions, Solifugae and spiders are generalist predators and compete for the same types of arthropod prey. At the same time, they frequently eat one another (intraguild predation). Interactions within natural communities are always complex, and the combined effects of competition and predation often engender concurrent differences in the dynamics, growth, survival and use of resources within age classes of interacting populations. Intraguild predation is a ubiquitous and often powerful interaction central to the structure and functioning of many natural communities (Polis et al. 1989; Sect. 7.7.2). The specificity of the micro-habitat in burrows and beneath rocks occupied by desert scorpions is associated with ecobiomes that have been constant for millions of years. Consequently, evolution has taken place primarily in terms of their biotic interactions, and life histories have developed only qualitative traits.

Similar ecological patterns can be caused by different processes; in order to cast light on the interrelations of the different taxa, Polis and McCormick (1986) removed more than 6000 scorpions (*Paruroctonus mesaenses*) from 300 (10 x 10 m) quadrats in a 29-month experiment in the Californian desert. Significantly more spiders, but not Solifugae, appeared in the experimental quadrats from which scorpions had been removed than in the control quadrats. Two alternative hypotheses might explain this result: exploitation competition for jointly exploited prey or intraguild predation. There was no evidence of the former, and no differences were found between experimental and control quadrats in prey abundance, or in the size and reproductive characteristics of the spiders. It therefore appears that release from predatory pressure in the areas from which scorpions were removed must have produced the increase in abundance of spiders. Moreover, these differences were apparent in spring, but not in the autumn. This was probably due to the fact that spiders usually mate in spring, at which time the adults are larger, more numerous, and move about more. The absence of any noticeable effect on the solifugid population, although these are eaten by scorpions more frequently than spiders, may have been due to the fact that most solfugids move continually and travel long distances, so that it was not possible to detect any changes resulting from the removal of scorpions. Finally, natural systems are much more complex than theory predicts; resource partitioning is characteristic of guilds; predation occurs widely among exploitative competitors, and intraguild predation may be a major mortality factor, producing differential patterns of resource use among guild members.

Competition among desert invertebrates has been reviewed by Heatwole (1996) and will not be considered further here.

The ecological variables that control the diversity of desert lizards in North America, southern Africa and Australia have been studied in great detail by Pianka (1975, 1986). Pianka concluded that, for the deserts of the south-western United States, most variation in lizard species diversity is determined by the causal chain – climate affects vegetation structure which, in turn, influences lizard diversity. The Australian deserts are the richest in the world in number of lizard species. As many as 40 different species can be found together in these deserts, four times the North American maximum. Similarly, habitats in the Kalahari desert support 11–18 species of lizards, about twice the number in North America.

Desert lizards subdivide their resources in three major ways, according to what they eat, where they forage, and when they are active. When these three niche dimensions were analysed, niche breadths were found to be almost identical in the three desert areas. Surprisingly, niche overlap was negatively related to species diversity, which may indicate strong competition, but neither niche breadth nor overlap explain the high diversity of species found in Australia. Apparently more lizard species are found in the Australian deserts because they exploit more resources. Consequently, total niche space is larger in Australia than in southern Africa and North America, where there are more ground-living insectivorous birds. Competition is therefore lower in Australia, and lizards there usurp the ecological roles played by other taxonomic groups in southern Africa and North America. Community interactions occur on a broad framework involving different taxonomic groups, and diversity within the whole community, rather than in taxonomic subdivisions, really needs to be studied. This point qualifies some of the conclusions reached earlier (Chap. 1).

The important idea that emanates from Pianka's work, as Krebs (1985) points out, is that we must look at community interactions on a broad framework because competition may be occurring between quite different taxonomic groups, such as birds and lizards. Taxonomic relationship does not imply any ecological interrelationship. The diversity in a whole community needs to be studied rather than diversity within its systematic subdivisions. Interspecific interactions between reptiles are discussed in detail by Heatwole and Taylor (1987).

The outcome of interspecific competition may depend upon exceptional, specific circumstances. For instance, in an experiment carried out in the Carmel Valley, California, the golden-crowned sparrow (*Zonotrichia atricapella*: Emberizidae) was found to occupy an area providing seeds and succulent shoots, in consequence of which it had no need to drink. Juncos (*Junco hyemalis*: Emberizidae) were found nearby, but excluded from this particular region by sparrows. When these were removed by trapping, the sparrows' area was occupied by juncos. The latter eat only seeds and therefore need a supply of liquid water. In the study area, however, there was a trough of water from which they had previously been excluded by golden-crowned sparrows.

Interspecific competition is manifested more frequently among desert mammals than it is in desert birds and, of the former, especially in rodents. Interspecific competition among rodents has been reviewed by Grant (1972), who showed that a species establishing dominance in one type of habitat may not succeed in another. The outcome in such cases is settled before there is any important competition for food, or any depression of fecundity. Several mechanisms of co-existence contribute to the diversity of granivorous rodent communities in arid lands. In the case of the sympatric North American *Perognathus amplus, Dipodomys merriami* and *Spermophilus tereticaudus*, seasonal rotation in foraging efficiencies makes co-existence possible if there is a trade-off between the costs of foraging in the different species. The species most efficient at foraging changes seasonally. Spatial variation in resource abundance makes co-existence possible if there is a trade-off between foraging efficiency and the cost of travel – which explains the presence of *Ammospermophilus harrisii* in the community. This species apparently prefers to forage at a large number of widely spaced patches to a high "giving-up" density, rather than to forage a smaller number of patches to a lower "giving-up" density (J. S. Brown 1989a).

Abramsky and Pinshow (1989) investigated changes in the foraging efforts of two gerbil species in respect to habitat type, intraspecific and interspecific activity in the Negev desert. *Gerbillus allenbyi* (mean mass

26 g) and *G. pyramidum* (mean mass 40 g) were studied in six 1-ha enclosures, and it was found that per capita foraging and non-foraging activity were negatively and significantly correlated with population density in both species. When *G. pyramidum* was common, *G. allenbyi* limited its activity to stabilized sand while *G. pyramidum* was active in semi-stabilized dunes. When none or only one *G. pyramidum* was present, the other gerbils (*G. allenbyi*) transferred their activity to the semi-stabilized dunes. Per capita activity of each species declined with increasing activity of the other. The experiment was continued by Abramsky et al. (1990), who concluded that (1) both species prefer the same habitat, and (2) exhibit intraspecific density-dependent habitat selection: (3) habitat density of each species was affected by the density of the other, and (4) habitat preferences of the two collapse from a centrifugal to a shared-preference model of habitat selection.

Similar results have been obtained in studies of the rodent faunas of other desert regions. For example, the nocturnal hairy-footed dune gerbil (*Gerbillurus tytonis*) and the day-active striped mouse (*Rhabdomys pumilio*) occur together in vegetative islands in the dune sea of the Namib desert and show a marked preference for the same micro-habitat. They live beneath bushes of nara (*Acanthosicyos horridus*; Fig. 32) which afford protection from predation (Sect. 6.4). The gerbils also inhabit a second micro-habitat where the risk of predation is greater: they increase their foraging at full moon when there is better illumination. Striped mice are more dependent upon nara bushes than are gerbils, and tend to form runways beneath them. It was found that removal of striped mice resulted in a significant increase in foraging activity by the hairy-footed gerbils. High susceptibility to predation apparently results in the shared preference for the safest habitat, despite competition for limited food resources. This is achieved by the temporal segregation of activity (Hughes et al. 1994).

Earlier work on environmental heterogeneity and the co-existence of desert rodents has been reviewed by Kotler and J. S. Brown (1988), while the mechanisms underlying community organization have been reviewed by J. S. Brown (1989b) in terms of species' trade-offs in the Negev. The situation is always more complicated when predators are present. Kotler et al. (1993) investigated the influence of diadema snakes (*Spalerosophis diadema*: Colubridae) on the gerbils *G. allenbyi* and *G. pyramidum*. They found that when snakes were present, *G. pyramidum* foraged fewer resource patches in the absence of added illumination at night. They preferred open habitats and may have intensified their use in the presence of snakes. In contrast, *G. allenbyi* intensified its use of bush micro-habitats in the presence of added illumination. Thus, there was a shift in the use of resource patches in response to the risk of predation. Moreover, the presence of one predator may make it easier for another to catch the prey – this is known as predator facilitation.

Predation by owls is a greater risk to gerbils in open micro-habitats: in closed micro-habitats, snakes are the greater danger. In this paper, Kotler et al. (1993) summarize their earlier work and that of their colleagues on the risks of predation to desert rodents.

Turning to larger mammals, McNaughton and Georgiadis (1986) have reviewed the ecology of African grazing and browsing mammals in both moist dystrophic and arid eutrophic savannas in respect of feeding ecology, environmental heterogeneity, body size and adaptive radiation. Almost nothing is known of population dynamics or the interactions between species. Some information is, however, available about the interactions between different species of carnivores. The geographical ranges of three extant species of jackal, the side-striped (*Canis adusta*), golden (*C. aureus*) and black-backed (*C. mesomelas*), overlap only in East Africa. Elsewhere, each species exists alone or in sympatry with only one other species of jackal. There is no evidence of size divergence between the three species in East Africa but, on the contrary, they appear to converge on a similar size. Divergence in shape, however, is present, although the evidence for it is subtle. The recent insinuation of golden jackals into the East African fauna has resulted in character divergence and probable niche compression of black-backed jackals, as shown by adaptations for increased carnivory, reduced sexual dimorphism and reduced variability. Neither species, however, has diverged functionally. The golden jackal shows dietary adaptations intermediate between those of the other two species, but somewhat more similar to those of the black-backed jackal. This species exhibits a selection for increased carnivory, which may have been favoured by the abundance of vertebrate prey in the plains of East Africa (Van Valkenburgh and Wayne 1994). At the same time, the hunting success of carnivores such as leopards is affected by the extent of vegetative cover (Bothma et al. 1994).

Interspecific competition among predatory species is often reduced or eliminated by ecological separation, in both their times of activity and their preferred environments. Thus, lions in the Serengeti tend to be found in woodland, hyaenas and cheetahs on the plains and borders of woodland, leopards in thickets and so on. Cheetahs and wild dogs are day-active, while the other large mammalian predators are mainly nocturnal (Schaller 1972). Again, striped hyaenas are solitary and more omnivorous than are spotted hyaenas, scavenging a great deal and eating insects, fruit and small vertebrate prey. The diets of the two species have several items in common, however, and there is competition where their ranges and habitats overlap (Kruuk 1976). Interactions between animals in arid land ecosystems have been reviewed by Wagner and Graetz (1981).

7.7.4 Effects of Herbivory on Plant Populations

The effects of herbivory on individual plants has been discussed above in some detail (Chap. 6) and mention made of population effects. In the present section, attention will be focussed briefly upon its effects on the structures of plant communities. We have seen that there is no compelling evidence to show that herbivory increases plant fitness, but selective grazing upon one species may benefit another in a competitive situation even if it, too, is eaten to some extent by the same herbivore (Sect. 7.7.1). In such a way, the relative proportions of the two species in the community may be influenced. From this point of view, deserts present an unusual situation. Perennial plants tend to be relatively scarce – the condition in which herbivory exerts a maximum effect. At the same time, they are clumped in scattered drainage areas and wadi beds where they are protected by their inaccessibility (Fig. 26). These categories coincide with Monod's (1954) modes "contracté" and "diffus" of the vegetation of the Sahara.

The selective effects of insects can influence entire ecosystems in the relatively simple environment of dune sands which, by their very nature, tend to be arid. Thus, in North America, the flea beetle *Altica supplicata* (Chrysomelidae) influences the growth and survival of its host plant, the sand dune willow (*Salix cordata*: Salicaceae) and consequently the pattern of plant succession on sand dunes. Experimentally protected plants added 2.2 times as much height, and 2.0 times as much diameter, as did plants exposed to beetles. Mortality was also strongly affected by the exclusion treatment (Bach 1994).

Plant populations in arid shrublands likewise are highly responsive to the effects of both climate and grazing. Changes are, however, less obvious and more gradual than in mesic areas, but nevertheless may be profound. The responses of the various species to grazing depend not only upon their palatability to livestock but also upon their life histories and physiological characteristics. Few studies have attempted to unravel the effects of grazing versus drought, although it is generally believed that grazing accentuates the effects of drought or that the response of a species to grazing during drought is predictable in terms of its past response to grazing. Over a 2-year period, Chambers and Norton (1993) found that the responses of desert shrubs in Utah were more predictable from their life history and physiological traits than from past responses to grazing alone. Light to moderate grazing and the removal of livestock before the active physiological growth of cool-season species had the least negative effects on population dynamics during a period of drought. The grazing regime increased survival or natality of certain species.

In prairie and steppe land, rodents usually consume less than 1% of annual primary production, but Chew and Chew (1970) reported that

9.5% was removed in south-eastern Arizona by *Dipodomys merriami, Anychomys torridus* and *Lepus californica* in a creosote bush community, while Scholt (1973) found that as much as 10.7% was removed by *D. merriami* in the Mojave desert. In the winter quarter, 75% of seeds and 35% of green vegetation were consumed. In 1970, more than 95% of the estimated total production of *Erodium cicutarium* (Geraniaceae) was taken, and 90% of total seeds eaten. This was enough to reduce the density of the plants by more than 30% in 1971, and Scholt concluded that the region was on the brink of over-exploitation. He also considered that herbivores are, in general, limited by their food supply in deserts, where productivity is so variable. If the green parts of plants are eaten selectively, the level of herbivory could have a large impact on total primary production.

Population cycles among herbivores are caused by interactions between them and their plant food – not by predator-prey, parasite-host or plant-environment interactions. Herbivore populations in deserts, as everywhere else, are limited by the availability of high-quality food. If they were limited by predation, any increase in plant productivity would engender an increase in the abundance of predators, thus leaving the herbivore equilibrium unaltered (Crawley 1983). Invertebrate predators that are regulated by natural enemies at low densities have virtually no impact on plant dynamics when re-growth or other forms of compensation occur. Interactions between plants and animals in arid land ecosystems have been reviewed by Graetz (1981).

7.8 Food Webs

Although there are seldom more than five major links in the food chains that interact to form any food web, complexity is introduced by the numbers of species and individuals within the web. Even in deserts, which are regarded as comparatively simple ecosystems, food webs may be extremely complicated (Cloudsley-Thompson 1974b). Louw and Seely (1982) compiled a generalized food web for the Namib desert, based on information in Seely and Louw (1980), but, because of the extensive overlap in feeding activities of the fauna, only one major food web, ending in *Bitis peringueyi*, could be indicated. The wide trophic ratios established between animal biomass and the biomass of plants and detritus (1:249) suggest that available energy does not limit the size of herbivore and omnivore populations. In contrast, the low protein content of the vegetation during dry periods (3.1%) is an important limiting factor. Unique features of the ecosystem are seen in the absence of a significant microbial decomposition loop, the degradation of large amounts of detritus accumulated by tenebrionid beetles, the dependence of the biota on fog, and the physiological and behavioral characteristics of the fauna. Average species richness of detritivores and carnivores is similar in the Namib dunefield, but individual numbers and estimated biomass are much greater for detritivores (C. S. Crawford and Seely 1987).

A more recent food web has been constructed for the coastal dune fauna of the Namib by McLachlan (1991; Fig. 37). This author concluded that dune food chains are fuelled by autochthonous primary production of the dune flora and organic inputs from the sea and land. He distin-

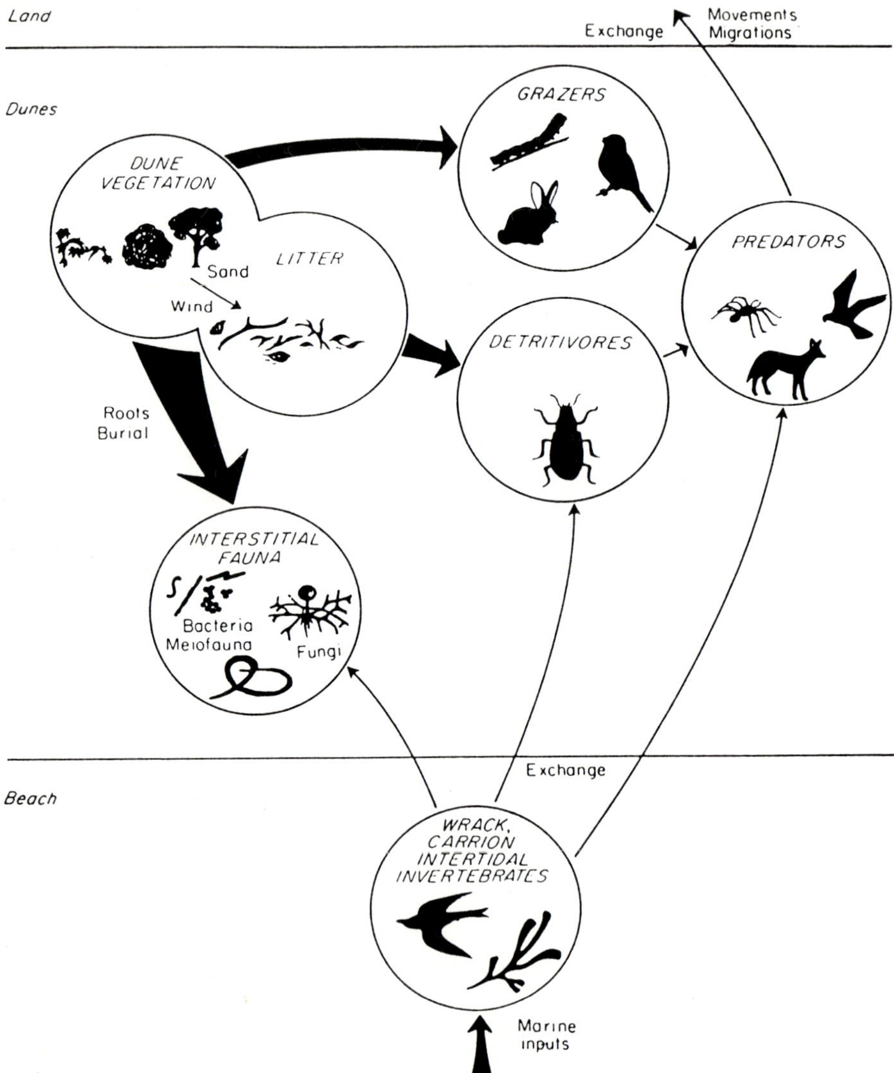

Fig. 37. The trophic structure of coastal desert dunes including three food chains and exchanges. (McLachlan 1991)

guished three food chains – a grazing food chain, which includes herbivorous insects, mammals, seed-eating and fruit-eating bats; a detrital food chain, occupied by macroscopic detritivores, particularly insects; and an interstitial food chain in the sand. This is composed of bacteria, actinomycetes, algae, fungi and meiofauna. The first two chains may be linked by common predators. The third chain, probably the most important, is largely separate. A prominent feature of coastal dunes, in contrast to inland desert dunes, may be the greater importance of the interstitial pathway. This is because the moist conditions and large quantities of buried organic matter favour a rich fauna of bacteria, fungi and meiofauna. Soil communities in deserts are discussed, among others by Wallwork (1982) and, more recently, by C. S. Crawford (1991b) and Zak and Freckman (1991).

In contrast to such complex webs, isolated food chains may also occur, on an extremely small scale, in arid environments. For example, fog and dew in the Namib desert provide water for the so-called Fensterpflanzen or window-algae which grow on the undersurfaces of translucent quartz rocks and stones. Light passing through the quartz is the limiting factor for photosynthesis and the result is often a ring of algae around the stone, several cm deep. Each stone represents a microcosm of plants (algae), grazers (Collembola and Psocoptera) and predators (mites and pseudoscorpions; Seely 1987). When parasites are taken into account (Sect. 7.2), food webs become enormously complicated, and maximum chain lengths increase as increasing numbers of species are involved. Chain lengths, longer than expected when compared with theoretical predictions, were noted in the Coachella Valley desert of California (Polis 1991c). Food web theory is not descriptive or predictive of nature there. Complexity arises from the large number of interactive species, the frequency of omnivory, age structures, looping, lack of compartmentalization and the diversity of the arthropod soil fauna. The same no doubt applies to other desert areas.

Food webs in desert communities have also been discussed in some detail by Polis (1991b) and by other authors in Polis (1991a). Naturally, they are less complex than in mesic regions, where the diversity of flora and fauna is greater. Nevertheless, they are very much more complicated than has been realised until comparatively recent times.

8 Discussion and Conclusions

In an inspiring introduction to elementary biology, Sir Arthur Shipley (1923. *Life*. Cambridge Univ Press, Cambridge) pointed out that the essence of animal life can be summed up in the conjunction of the French verb manger: Je mange, tu manges, il mange etc., or its terrible correlative, Je suis mangé, tu es mangé .... Field workers throughout the world, and especially in deserts – for reasons constantly alluded to in this book – will not doubt the validity of this conclusion. Parasitism, disease and starvation may be important factors in controlling animal populations, but even weak and ailing animals are frequently killed by predators before death claims them from some other cause. Various additional forms of biotic interaction have been discussed in the later chapters, as they apply to the plants and animals of arid lands. In this final chapter, an attempt will be made to identify certain general ecological principles that apply to all desert organisms, and thereby to extend the discussions on evolutionary parallels (Sect. 1.1) and ecological analogues (Sect. 1.2) with which the first chapter was largely concerned.

Deserts comprise one of the world's major terrestrial ecobiomes. Apart from a few pioneering works, such as those of Buxton (1923) and Kachkarov and Korovine (1942), they were relatively little studied before the 1950s. Most of the early work was descriptive and autecological: more recently, however, the emphasis has turned towards synecological and experimental studies, summarized, for instance, in the books of Whitford (1986a), Schmidt (1990) and Polis (1991a). In the present book I have tried to provide some interconnections between the more strictly autoecological and synecological volumes in the current Series including especially those by Gutterman (1993), G. Costa (1995), Sømme (1995), Heatwole (1996) and others forthcoming, which deal with more general concepts. No attempt has been made, however, to consider the theoretical or mathematical aspects of evolutionary strategies (Stearns 1976).

8.1 Emerging Principles

In the study of the biotic interrelationships of desert plants and animals, certain obvious - but sometimes overlooked - basic principles emerge. These will now be mentioned briefly.

8.1.1 Multiple Functions

Some of the issues over which biologists disagree are concerned with function. Structural, physiological and behavioral characters may serve multiple functions: advocates of one are not necessarily wrong when another is established as correct. Transpiration not only cools the leaves of a plant, but it may serve to eliminate surplus water. Moreover, as already mentioned (Sect. 7.7.1), it can be a factor in competition with nearby plant associates, by denying them water when it is available. Numerous contradictory functions have been attributed to the pectines of scorpions (Fig. 38; Cloudsley-Thompson 1955). Some of the earlier hypotheses were quite bizarre; but claims that pectines serve to detect ground vibrations caused by the footfalls of large avian and mammalian predators, that they are used to determine the nature of the substrate for the selection of suitable places for the deposition of spermotophores, and that the functions of their peg sensilla are simultaneously tactile, proprioceptive, moisture receptive and chemoreceptive (Root 1990) cannot be dismissed lightly. In my opinion, none of these possible functions is incompatible with any other one.

Gabar goshawks (*Micronisus gabar*) incorporate large nests of social spiders (*Stegodyphus* spp.) into their own nests. The spiders continue to live in and around the birds' nests about 3-12 m above the ground, which is well above the normal height (1 m) of *Stegodyphus* webs. This unusual association can be explained in various ways: (1) the spider nest is incorporated as a lining or padding and the activities of the spiders are incidential; (2) the large spider webs camouflage the birds' nests or nestlings from predators; (3) the spiders' silk strengthens the structure of the goshawk's nest; (4) the spiders check the activities of insect scavengers, parasitic flics and other pests of juvenile goshawks; (5) spiders are brought to the bird's nest as food, and some manage to survive and re-establish themselves in the new locality (Henschel et al. 1991). Which of these is the correct explanation or explanations has yet to be determined, but they are not mutually exclusive, and empirical evidence is lacking. Several are probably involved. The dark chanting goshawk (*Melierax metabates*) has also been reported to incorporate social spider nests into its own nest, but neither the pale chanting goshawk (*M. canorus*) nor other small hawks do so.

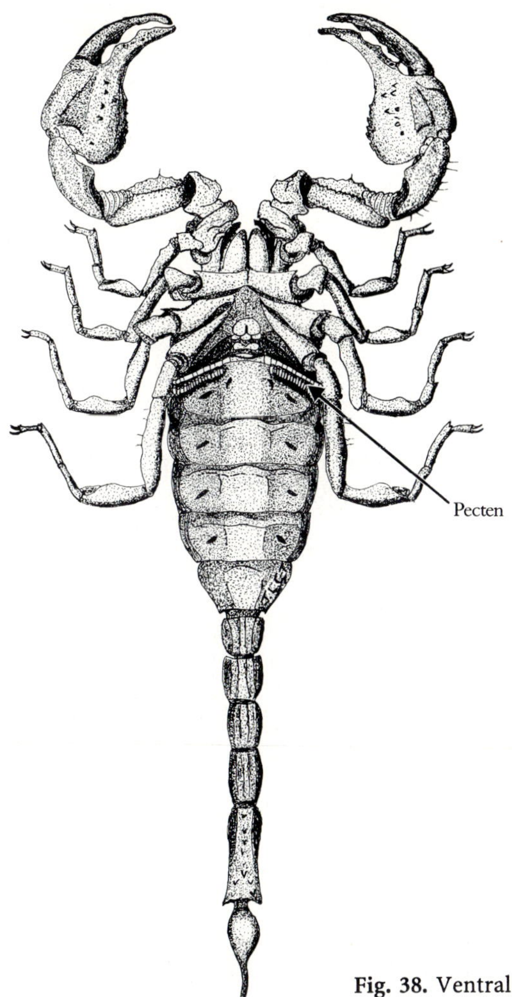

Fig. 38. Ventral view of a scorpion, showing pectines. (Cloudsley-Thompson 1980)

The stripes of zebras [*Eguus* (= *Hippotragus*) spp.] may appear cryptic from a distance at certain times of day; they may be conspicuous at close range so that a predator gives chase too soon and its prey escapes; they may reinforce the bond between the sexes; and they may afford protection from tsetse (*Glossina* spp.) and other biting flies. Quite possibly, all these hypotheses are valid.

Anti-predator behaviour patterns of ungulates include pronking by springbok, tail-flagging by white-tailed deer (*Odocoileus virginianus*) and stotting by Thomson's gazelles (Sect. 4.1). These not only warn conspecifics of danger, but inform predators that they have been detected. The behaviour of prey depends very much upon that of their predators, but little attempt has yet been made to relate the distribution of the various types of anti-predator behaviour in different prey species, to their habitats, grouping patterns and the principle species of predators that attack them. Although defence is implicit, little is known as to how the defences of a single prey species operate against suites of predators, each hunting in different ways (Crawley 1992). Doubtless, all such responses have multiple functions.

8.1.2 Selective Compromise

Some functions are indeed incompatible. For example, most desert reptiles are cryptic but, when the ambient temperature increases above a specific threshold, the skin blanches, and reflection of solar radiation increases. This reduces overheating, but it makes the reptile more conspicuous to its predatory enemies. Colour change in reptiles is correlated with both concealment and thermoregulation; but the operation of the two is largely synergistic. When they produce opposing effects, the one possessing the greater survival value, at that moment, is selected (Cloudsley-Thompson 1979, 1991). Antelope horns originally evolved for territorial disputes, but are also effective deterrance against predatory enemies. Although their owners may be cryptically coloured, the horns are dark, which draws attention to them as weapons of defence (Sect. 4.8). There is often a trade-off between one function and another, resulting in the selection of a compromise.

8.1.3 Temporal Application of Adaptations

A particular adaptation needs to be effective only for brief periods in order to be selected. Thus, a bird may require to be cryptic only during the breeding season when it is incubating the eggs; but it maintains its plumage colours throughout the rest of the year. The mane of a male lion may occasionally be used in threat display to rival lions but, at other times, it has little known function except, possibly crypsis (Sect. 7.1.4). Indeed, it may even increase the animal's thermal load, thereby invoking selective compromise (Sect. 8.1.2). Again, at certain times of day only is the zebra's coloration likely to be cryptic (Sect. 8.1.1). Nevertheless, selection operates continually, impressing its lasting influence upon all living organisms.

8.1.4 Competitive Advantages

Plants and animals tend to be found, not necessarily in the environment most suited to their climatic and nutritional requirements, but in places where they compete best with other species. For example, addax antelope are relatively defenceless in comparison with oryx, yet they are able to survive without drinking water in the most arid regions of the Sahara where even oryx cannot live. Here, they are relatively inaccessible to carnivorous predators against which they would be unable to defend themselves – even though oryx might be able to do so (Sect. 4.8). Addax would probably grow faster and be more healthy, living in a less rigorous habitat, but there they might compete unfavourably with oryx and other antelope less well adapted to aridity, in adddition to being exposed to predation by lions and leopards. As we have seen (Sect. 7.7.3), a rodent species may be dominant in one type of habitat but not in another (Grant 1972). Another example is afforded by tadpole shrimps (*Triops* spp.: Branchiopoda) and other crustaceans that are restricted to temporary rainpools in the desert where their entire life spans are compressed into a couple of weeks; but they thereby avoid predation from fishes (Cloudsley-Thompson 1991).

8.1.5 Diminishing Returns

Perfection is energetically costly. The degree to which a mimic has evolved to resemble its model must logically depend upon the extent to which it is vulnerable to predation. Moths with cryptic coloration that disguises them as bark or lichen (Cott 1940) are almost invariably more symmetrical than the objects to which they show protective resemblance. Presumably, asymmetry is genetically more costly than is symmetry, although the former is by no means rare in nature. Throughout Chapter 7, mention has been made of trade-offs between the cost of one adaptation and another. These trade-offs involve both selective compromises and the concept of diminishing returns for metabolic expenditure.

8.1.6 Evolutionary Lability

Different taxa are by no means equal in their evolutionary lability – in addition to the obvious fact that *r*-selected species with short life cycles must inevitably evolve and adapt more rapidly than do K-selected species with longer life cycles – the latter may sometimes be more successful in less equable environments (Sect. 7.1). Moreover, taxonomic affinities shape community function, as well as the distribution and abundance of species (Case 1994).

Certain animal taxa are especially successful in arid environments, on account not so much of specific adaptations as of the innate qualities which pre-adapt them to life in hot, dry regions. These taxa include large arachnids – especially scorpions, Solifugae and mygalomorph spiders (Orthognatha) – large tenebrionid beetles, lizards and, to a lesser extent, small rodents. Mammals do not enjoy the advantage of ectothermy and the consequent reduction of nutritional requirements when their maintenance would lower lifetime reproductive success (Schall and Pianka 1978). Apart from some Tenebrionidae, they all tend to avoid the extremes of the midday heat behaviorally by escaping into shelters or deep burrows, while conservation of water is primarily physiological (Cloudsley-Thompson 1991). Their success, too, depends upon their interactions with other taxa and, indirectly, with desert plants. Furthermore, they tend to be extremely adaptable, both physiologically and behaviorally.

A species, the size of whose populations is limited by predation, would be expected to show greater dietary preference than one that is limited by its food resources. Co-existence of herbivorous species in any one habitat dictates certain combinations of specialists and generalists which, in turn, are strongly influenced by the turnover rates of the various populations. Australian seed-eating termites and ants evidently live well within their available food resources, but they are not dietary generalists at all seasons. Instead, they are flexibly selective, have a wide niche breadth, and concentrate on freshly fallen seed. The lack of competitive pressure between species may buffer ant populations from environmental fluctuations. The niche characteristics of small mammalian herbivores differ from those of harvester ants. Although termites show dietary flexibility, interspecific competition imposes constraints, while partitioning of food resources is determined by both the size of the seeds on which the termites feed, and the mammals that in turn prey on them – along with the mobility and territorial defence of the latter (Graetz 1981). Population structure can also influence the outcome of natural selection whenever individuals interact, ecologicaly or behaviorally, with a limited number of conspecifics relative to the total population.

8.2 Desert Ecosystems

Ecosystems have been defined in several ways. This book has been concerned with all the metazoan organisms that interact with one another in arid lands. Primary production is low there, due to water shortage, and the rate of accumulation of biomass is therefore also slow (Ludwig 1986). Upon the vegetation (McCleary 1968) directly depend herbivorous invertebrates (C. S. Crawford 1986) and vertebrates (J. H. Brown 1986) and, indirectly, their predators and parasites. Finally, decomposers – not only micro-organisms but the entire complex of detritivores, including beetles, ants, millipedes and so on, with their gut symbionts – in association with the soil mesofauna (Wallwork 1982) – are responsible for nutrient cycling (Whitford 1986b). They also concentrate the potential energy in detritus for the benefit of their predators!

At the present stage of knowledge, the qualitative aspects of many of these processes are reasonably well understood by inference, but quantitative studies are extremely scarce. For instance, interactions between rodents and ants in the Chihuahuan desert (Valone et al. 1994) are now known to be much more complex than those reported for the Sonoran desert some years earlier (J. H. Brown and Davidson 1977). Major interactions are mediated indirectly through the vegetation (Samson et al. 1992) rather than through direct competition for seeds. It is reasonable, therefore, to predict that much of future research on desert organisms will be concentrated upon experimental and quantitative population studies.

Although arid regions all experience shortage of water and other climatic characteristics that induce evolutionary parallels and analogues, subtle differences are present which result in the development of markedly dissimilar ecosystems in the various deserts of the world. In the Australian arid zone, for example, the species richness of ants is greater, and that of mammalian granivores less, than in the North American deserts. This is not due to paucity of rodents, however, and it appears that ecological differences between the two arid environments have had significantly different consequences for the assembly of animal communities (Morton and Davidson 1988). The same conclusion was reached by Pianka (1973) and by Schall and Pianka (1978) in the case of lizard communities. More recently, Winemiller and Pianka (1990) concluded that diverse lizard assemblages of the Australian deserts show significant partitioning of micro-habitats among ecologically similar species. Guild structure disappears and niche segregation becomes evident when more refined analytical methods are used. In the less diverse Kalahari desert, assemblages of lizards are more variable and not so tightly organized as they are in Australia.

The biotic relationships discussed in this book are widespread, and none of them is unique to arid lands. Moreover, by no means every type of intraspecific or interspecific relationship has been mentioned, although I have tried to include the most important ones. One aspect of social behaviour not discussed is play among mammals. Clearly this has measurable costs, so there must be a trade-off and benefits also, but the function of play is still somewhat uncertain (Fagen 1981). According to Caro (1995), the overall cost of play in young cheetahs is low, while it both teaches them to hunt and provides experience for escaping from predators. Whether desert species play less than related forms in more mesic environments has apparently not been investigated.

One of my intentions has been to draw attention to examples of biotic interactions, which although well known to biologists who are more familiar with the mesic regions of the world, have not yet been investigated in equal depth in the arid zones. I hope that this book will be of some use as an introduction to current literature in addition, perhaps, to achieving the other objects stated in the Introduction. Nevertheless, as Field-Marshall Helmuth von Moltke (1800–1891) wrote, no plan survives contact with the enemy! At the same time, in the words of the Emperor Marcus Aurelius, written 17 centuries earlier: "if I do err I will most gladly retract. For it is the truth that I seek after by which I am sure that no man was hurt, and equally sure that he is hurt who continues in ignorance and error."

References

(The literature survey was completed in the Spring of 1995)

Abdullah MAR, Abulfatih HA (1995) Predation of *Acacia* seeds by bruchid beetles and its relation to altitudinal gradient in south-western Saudi Arabia. J Arid Environ 29:99–105

Abramsky Z, Pinshow B (1989) Changes in foraging effort in two gerbil species with habitat type and intra- and interspecific activity. Oikos 56:43–53

Abramsky Z, Rosenzweig ML, Pinshow BP, Brown JS, Kotler BP, Mitchell WA (1990) Habitat selection: an experimental field test with two gerbil species. Ecology 71:2358–2369

Abramsky Z, Shachak M, Subrach A, Brand S, Alfia H (1992) Predator-prey relationships: rodent-snail interaction in the Central Negev Desert of Israel. Oikos 65:128–133

Abushama FT (1972) The repugnatorial gland of the grasshopper *Poecilocerus hieroglyphicus* (Klug). J Entomol Ser A Gen Entomol 47:95–100

Abushama FT (1984) Epigeal insects. In: Cloudsley-Thompson JL (ed) Sahara desert (Key environments). Pergamon Press, Oxford, pp 129–144

Alexander AJ (1958) On the stridulation of scorpions. Behaviour 12:339–352

Alexander AJ (1960) A note on the evolution of stridulation within the family Scorpionidae. Proc Zool Soc Lond 133:391–399

Alexander RD (1974) The evolution of social behaviour. Annu Rev Ecol Syst 5:325–383

Althoff DM, Thompson JN (1994) The effects of tail autotomy on survivorship and body growth of *Uta stansburiana* under conditions of high mortality. Oecologia 100:250–255

Applin DG, Cloudsley-Thompson JL, Constantinou C (1987) Molecular and physiological mechanisms in chronobiology – their manifestations in the desert ecosystem. J Arid Environ 13:187–197

Arnold EN (1984) Evolutionary aspects of tail shedding in lizards and their relatives. J Nat Hist 18:127–169

Arnold EN (1986) Mite pockets of lizards, a possible means of reducing damage by ectoparasites. Biol J Linn Soc 29:1–21

Arnold EN (1988) Caudal autotomy as a defense. In: Gans C, Huey RB (eds) Biology of the Reptilia 16 (Ecology B) Defense and life history. Alan R Liss, New York, pp 235–273

Arnold EN (1995) Identifying the effects of history on adaptation: origins of different sand-diving techniques in lizards. J Zool Lond 253:351–388

Askew RR (1971) Parasitic insects. Heinemann Educ Books, London

Augner M, Fagerström T, Tuomi J (1991) Competition, defense and games between plants. Behav Ecol Sociobiol 29:231–234

Bach CE (1994) Effects of a specialist herbivore (*Altica supplicata*) on *Salix cordata* and sand dune succession. Ecol Monogr 64:423–445

Baker BH (1960) Mammals of the Guadiana lava field. Durango, Mexico. Publ Michigan State Mus Biol Ser 1:305–327

Baker HG, Harris BJ (1957) The pollination of *Parkia* by bats and its attendant evolutionary problems. Evolution 11:449–460

Ballinger RE, Tinkle DW (1979) On the cost of tail regeneration to body growth in lizards. J Herpetol 13:374–375

Barbour MG (1981) Plant-plant interactions. In: Goodall DW, Perry RA, Howes KMW (eds) Arid-land ecosystems: structure, functioning and management (International Biological Programme 17), vol 2. Cambridge University Press, Cambridge, pp 33–49

Barlow B, Wiens D (1977) Host-parasite resemblance in Australian mistletoes: the case for cryptic mimicry. Evolution 31:69–84

Barth F (1985) Insects and flowers. The biology of a partnership (translated by Biederman-Thorson MA). George Allen & Unwin, London. Princeton University Press, Princeton

Barton NWH, Houston DC (1994) Morphological adaptation of the digestive tract in relation to feeding ecology of raptors. J Zool Lond 232:133–150

Batanouny KH (1981) Ecology and flora of Qatar. Alden Press, Oxford for Univ Qatar

Bates HW (1862) Contributions to an insect fauna of the Amazon valley. Lepidoptera: Heliconidae. Trans Linn Soc Lond 23:495–566

Bauer AM, Russell AP (1991) Pedal specializations in dune-dwelling geckos. J Arid Environ 20:43–62

Bayne EM, Brigham RM (1995) Prey selection and foraging constraints in common poorwills (*Phalaenoptilus nuttallii*: Aves:Caprimulgidae). J Zool Lond 235:1–8

Bazzaz FA, Chiariello NR, Coley PD, Pitelka LF (1987) Allocating resources to reproduction and defense. Bioscience 37:58–67

Bell EA (1981) The physiological roles of secondary (natural) products. In: Conn EE (ed) The biochemistry of plants, vol 7 Secondary plant products. Academic Press, New York, pp 1–19

Bell WJ (1991) Searching behaviour. The behavioural ecology of finding resources. Chapman & Hall, London

Bellairs A (1969) The life or reptiles (2 vols). Weidenfeld & Nicolson, London

Bellairs A, d'A, Bryant SV (1985) Autotomy and regeneration in reptiles. In: Gans C, Billett F (eds) Biology of the Reptilia 15 (Development B). Academic Press, London, pp 301–410

Belsky AJ, Carson WP, Jensen CL (1993) Over compensation by plants: herbivore optimization or red herring? Evol Ecol 7:109–121

Bennet EL, Bonner J (1953) Isolation of plant growth inhibitors from *Thamnosoma montana*. Am J Bot 40:29–33

Bennett NC (1994) Reproductive suppression in social *Cryptomys damarensis* colonies – a lifetime of socially induced sterility in males and females (Rodentia: Bathyergidae). J Zool Lond 234:25–39

Bennun L (1994) The contribution of helpers to feeding nestlings in grey-capped social weavers *Pseudonigrita arnaudi*. Anim Behav 47:1047–1056

Benson L (1969) The cacti of Arizona. University of Arizona Press, Tucson

Bertram BCR (1975) Living in groups: predators and prey. In: Krebs JR, Davies NB (eds) Behavioural ecology: an evolutionary approach. Blackwell, Oxford, pp 64–96

Bertram BCR (1992) The ostrich communal nesting system (Monographs in behavior and ecology). Princeton University Press, Princeton

Blum MS (1981) Chemical defences of arthropods. Academic Press, New York

Bonner J, Galston AW (1944) Toxic substances in the interaction of higher plants. Bot Gaz 106:185–198

Borremissza GF (1957) An analysis of arthropod succession in carrion and the effect of its decomposition on the soil fauna. Aust J Zool 5:1–12

Bothma JduP, van Rooyen N, Theron GK, le Riche EAN (1994) Quantifying woody plants as hunting cover for southern Kalahari leopards. J Arid Environ 26:273–280

Bowers MD (1990) Recycling plant natural products for insect defense. In: Evans DL, Schmidt JO (eds) Insect defenses. Adaptive mechanisms and strategies of prey and predators (SUNY series in animal behavior). State University of New York Press, Albany, pp 353–386

Bowers WS (1991) Insect hormones and antihormones in plants. In: Rosenthal GA, Berenbaum MR (eds) Herbivores. Their interactions with secondary plant metabolites, vol 1. The chemical participants, 2nd edn. Academic Press, San Diego, pp 431–456

Brahmachary RL (1982) The inter-relationship between *Acacia* thorns and insects in East Africa. J Arid Environ 5:319–322

Brain C (1990) Aspects of drinking by baboons (*Papio ursinus*) in a desert environment. In: Seely MK (ed) Namib ecology: 25 years of Namib research. Transvaal Museum Monograph No 7. Transvaal Museum, Pretoria, pp 169–172

Brattstrom BH (1974) The evolution of reptilian social behaviour. Am Zool 14:35–49

Brian MV (1983) Social insects. Ecology and behavioural biology. Chapman & Hall, London

Brisson J, Reynolds JF (1994) The effect of neighbors on root distribution in a creosote bush (*Larrea tridentata*) population. Ecology 75:1693–1702

Brown CR, Hoogland JL (1986) Risk in mobbing for solitary and colonial swallows. Anim Behav 34:1319–1323

Brown JH (1986) The roles of vertebrates in desert ecosystems. In: Whitford WG (ed) Pattern and process in desert ecosystems. Contributions of the Committee on Desert and Arid Zones Research of the Southwestern and Rocky Mountain Division of the American Association for the Advancement of Science 21. University of New Mexico Press, Albuquerque, pp 51–71

Brown JH, Davidson DW (1977) Competition between seed-eating rodents and ants in desert ecosystems. Science (NY) 196:880–882

Brown JH, Grover JJ, Davidson DW, Lieberman GA (1975) A preliminary study of seed predation in desert and montane habitats. Ecology 56:987–992

Brown JL (1987) Helping and communal breeding in birds: ecology and evolution. Princeton University Press, Princeton

Brown JS (1989a) Desert rodent-community structure: a test of four mechanisms of coexistence. Ecol Monogr 59:1–20

Brown JS (1989b) Mechanisms underlying the organization of a desert rodent community. J Arid Environ 17:211–218

Brown L (1976) Birds of prey, their biology and ecology. Hamlyn, London

Brown LH (1982) Ciconiidae, storks. In: Brown LH, Urban EK, Newman K (eds) The birds of Africa, vol 1. Academic Press, London, pp 172–190

Brown WL (1960) Ants, acacias and browsing mammals. Ecology 41:587–592

Brownell PH (1977) Compressional and surface waves in sand: used by desert scorpions to locate prey. Science (NY) 197:479–482

Brownell PH, Farley RD (1979a) Prey localizing behavior of the nocturnal desert scorpion *Paruroctonus mesaensis*: orientation to substrate vibrations. Anim Behav 27:185–193

Brownell PH, Farley RD (1979b) Detection of vibrations in sand by tarsal sense organs of the nocturnal scorpion, *Paruroctonus mesaensis*. J Comp Physiol 131A:23–30

Brownell PH, Farley RD (1979c) Orientation to vibrations in sand by the nocturnal scorpion *Paruroctonus mesaensis*: mechanism of target location. J Comp Physiol 131A:31–38

Bubenik GA (1990) Epigenetical, morphological, and behavioral aspects of evolution of horns, pronghorns and antlers. In: Bubenik GA, Bubenik AB (eds) Horns, pronghorns and antlers. Evolution, morphology, physiology and social significance. Springer, Berlin Heidelberg New York, pp 3–113

Bull CM, Burzacott D (1993) The impact of tick load on the fitness of their lizard hosts. Oecologia 96:415–419

Burrage BR (1973) Comparative ecology and behaviour of *Chamaeleo pumilis pumilis* (Gmelin) and *C. namaquensis* A. Smith (Sauria: Chamaeleonidae). Ann S Afr Mus 61:1–158

Buxton PA (1923) Animal life in deserts. A study of the fauna in relation to the environment. Edward Arnold, London

Carlisle DB, Ellis PE, Betts E (1965) The influence of aromatic shrubs in sexual maturation in the desert locust *Schistocerca gregaria*. J Insect Physiol 11:1541–1558

Caro TM (1995) Short-term costs and correlates of play in cheetahs. Anim Behav 49:333–345

Caro TM, FitzGibbon CD (1992) Large carnivores and their prey: the quick and the dead. In: Crawley MJ (ed) Natural enemies. The population biology of predators, parasites and diseases. Blackwell, Oxford, pp 117–142

Carpenter GDH (1941) The relative frequency of beak marks in butterflies of different edibility to birds. Proc Zool Soc Lond 111(A):223–231

Case T (1994) Population and community ecology. In: Vitt LJ, Pianka ER (eds) Lizard ecology. Historical and experimental perspectives. Princeton University Press, Princeton, pp 261–265

Chambers JC, Norton BE (1993) Effects of grazing and drought on population dynamics of salt desert shrub species on the Desert Experimental Range, Utah. J Arid Environ 24:261–275

Cheng TC (1986) General parasitology, 2nd edn. Academic Press College Division, Orlando

Chew RM, Chew AE (1970) Energy relationships of the mammals of a desert shrub (*Larrea tridentata*) community. Ecol Monogr 40:1–21

Chilton NB, Bull CM, Andrews RH (1992) Niche segregation in reptile ticks: attachment sites and reproductive success of females. Oecologia 90:255–259

Cloudsley-Thompson JL (1955) On the function of the pectines of scorpions. Ann Mag Nat Hist (12)9:556–560

Cloudsley-Thompson JL (1958) Spiders, scorpions, centipedes and mites. Pergamon, London (2nd edn 1968, Pergamon, Oxford)

Cloudsley-Thompson JL (1960a) A new sound-producing mechanism in centipedes. Entomol Mon Mag 96:110–113

Cloudsley-Thompson JL (1960b) Adaptive functions of circadian rhythms. Cold Spring Harbor Symp Quant Biol 25:345–355

Cloudsley-Thompson JL (1961a) Rhythmic activity in animal physiology and behaviour. Academic Press, New York

Cloudsley-Thompson JL (1961b) Observations on the natural history of the "camel spider" *Galeodes arabs* CL Koch (Solifugae: Galeodidae) in the Sudan. Entomol Mon Mag 97:145–152

Cloudsley-Thompson JL (1962) Some aspects of the physiology and behaviour of *Dinothrombium* (Acari). Entomol Exp Appl 5:69–73

Cloudsley-Thompson JL (1963) A note on the association between *Bengalia* spp. (Dipt., Calliphoridae) and ants in the Sudan. Entomol Mon Mag 98:177–179

Cloudsley-Thompson JL (1964) Terrestrial animals in dry heat: arthropods. In: Dill DB (ed) Handbook of physiology, Sect 4. American Physiological Society, Washington DC, pp 451–465

Cloudsley-Thompson JL (1965) Animal conflict and adaptation. GT Foulis, London

Cloudsley-Thompson JL (1968) The Merkhiyat jebels: a desert community. In: Brown GW Jr (ed) Desert biology, vol 1. Academic Press, New York, pp 1–20

Cloudsley-Thompson JL (1970) Recent work on the adaptive functions of circadian and seasonal rhythms in animals. J Interdiscip Cycle Res 1:5–19

Cloudsley-Thompson JL (1971) The temperature and water relations of reptiles. Merrow, Watford (Herts)

Cloudsley-Thompson JL (1974a) The ecology of oases. Merrow, Watford (Herts)

Cloudsley-Thompson JL (1974b) Desert life (The living earth). Aldus Books, London

Cloudsley-Thompson JL (1977a) Man and the biology of arid zones (Contemporary biology). Edward Arnold, London

Cloudsley-Thompson JL (1977b) Adaptational biology of the Solifugae (Solpugida). Bull Br Arachnol Soc 4:61–71

Cloudsley-Thompson JL (1979) Adaptive functions of the colours of desert animals. J Arid Environ 2:95–104

Cloudsley-Thompson JL (1980) Tooth and claw. Defensive strategies in the animal world. JM Dent, London

Cloudsley-Thompson JL (1981a) Bionomics of the rainbow lizard *Agama agama* (L.) in eastern Nigeria during the dry season. J Arid Environ 4:235–245

Cloudsley-Thompson JL (1981b) A comparison of rhythmic locomotory activity in tropical forest Arthropoda with that in desert species. J Arid Environ 4:327–334

Cloudsley-Thompson JL (1984a) Introduction. In: Cloudsley-Thompson JL (ed) Sahara desert (Key environments). Pergamon Press, Oxford, pp 1–15

Cloudsley-Thompson JL (1984b) Arachnids. In: Cloudsley-Thompson JL (ed) Sahara desert (Key environments). Pergamon Press, Oxford, pp 175–204

Cloudsley-Thompson JL (1986) Desert adaptations in arachnids. Proc 9th Int Congr Arachnol, Panama, 1983. Smithsonian Institution Press, Washington, DC, pp 29–32

Cloudsley-Thompson JL (1989) Behavioural adaptations of desert Arthropoda. Boll Accad Gioenia Sci Nat (1987) 20(332):283–293

Cloudsley-Thompson JL (1990) Thermal ecology and behaviour of *Physadesmia globosa* (Coleoptera: Tenebrionidae) in the Namib desert. J Arid Environ 19:317–324

Cloudsley-Thompson JL (1991) Ecophysiology of desert arthropods and reptiles (Adaptations of desert organisms). Springer, Berlin Heidelberg New York

Cloudsley-Thompson JL (1993a) The adaptational diversity of desert biota. Environ Conserv 20:227–231

Cloudsley-Thompson JL (1993b) Spiders and scorpions (Araneae and Scorpions). In: Lane RP, Crosskey RW (eds) Medical insects and arachnids. Chapman & Hall, London, pp 659–682

Cloudsley-Thompson JL (1994) Predation and defence amongst reptiles. R & A Publishing, Bristol

Cloudsley-Thompson JL (1995) A review of the anti-predator devices of spiders. Bull Br Arachnol Soc 10:81–96

Cloudsley-Thompson JL, Constantinou C (1984) Stridulatory apparatus of Solifugae (Solpugida). J Arid Environ 7:365–369

Cloudsley-Thompson JL, Lourenço WR (1994) The origin of desert faunas. Biogeographica 79:183–192

Cockburn A (1988) Social behaviour in fluctuating populations. Studies in behavioural adaptation. Croom Helm, London

Cody ML (1986) Spacing patterns in Mojave desert plant communities: near-neighbor analysis. J Arid Environ 11:199–217

Cody ML (1993) Do cholla cacti (*Opuntia* spp., subgenus *Cylindropuntia*) use or need nurse plants in the Mojave desert? J Arid Environ 24:139–154

Constantinou C, Cloudsley-Thompson JL (1984) Stridulatory structures in scorpions of the families Scorpionidae and Diplocentridae. J Arid Environ 7:359–364

Cooper SM (1990) The hunting behaviour of spotted hyaenas (*Crocuta crocuta*) in a region containing both sedentary and migratory populations of herbivores. Afr J Ecol 28:131–141

Cooper WE Jr (1990) Prey odor detection by teiid and lacertid lizards and the relationship of prey odor detection to foraging made in lizard families. Copeia 1990(1):237–242

Cooper WE Jr (1994a) Multiple functions of extraoral lingual behaviour in iguanian lizards: prey capture, grooming and swallowing, but not prey detection. Anim Behav 47:765–775

Cooper WE Jr (1994b) Prey chemical discrimination, foraging mode, and phylogeny. In: Vitt LJ, Pianka ER (eds) Lizard ecology. Historical and experimental perspectives. Princeton University Press, Princeton, pp 95–116

Costa G (1995) Behavioural adaptations of desert animals (Adaptations of desert organisms). Springer, Berlin Heidelberg New York

Costa G, Leonardi M-E, Petralia A (1987) Reproductive behaviour of the giant cricket *Brachytrupes membranaceus* (Drury) in the Namib. Madoqua 15(3):217–228

Costa M (1964) Descriptions of the hitherto unknown stages of *Parasitus copridis* Costa (Acari: Mesostigmata) with notes on its biology. J Linn Soc (Zool) 45:209–222

Cott HB (1940) Adaptive coloration in animals. Methuen, London

Cott HB (1947) The edibility of birds: illustrated by five years' experiments (1941–1946) on the food preferences of the hornet, cat and man: and considered with special reference to the theories of adaptive coloration. Proc Zool Soc Lond 116:371–524

Crawford CS (1981) Biology of desert invertebrates. Springer, Berlin Heidelberg New York

Crawford CS (1986) The role of invertebrates in desert ecosystems. In: Whitford WG (ed) Pattern and process in desert ecosystems. Contributions of the Committee on Desert and Arid Zones Research of the Southwestern and Rocky Mountain Division of the American Association for the Advancement of Science 21. University of New Mexico Press, Albuquerque, pp 73–91

Crawford CS (1991a) The community ecology of marco-arthropod detritivores. In: Polis GA (ed) The ecology of desert communities. University of Arizona Press, Tucson, pp 89–112

Crawford CS (1991b) Animal adaptations and ecological processes in desert dunefields. J Arid Environ 21:245–260

Crawford CS, Seely MK (1987) Assemblages of surface-active arthropods in the Namib dunefield and associated habitats. Rev Zool Afr 101:397–421

Crawford CS, Bercovitz K, Warburg MR (1987) Regional environments, life-history patterns, and habitat use of spirostreptid millipedes in arid regions. Zool J Linn Soc 89:63–88

Crawford RMM (1989) Studies in plant survival. Ecological case histories of plant adaptation to adversity. Studies in ecology, vol 11. Blackwell, Oxford

Crawley MJ (1983) Herbivory. The dynamics of animal-plant interactions. Studies in ecology, vol 10. Blackwell, Oxford

Crawley MJ (ed) (1992) Natural enemies. The population biology of predators, parasites and diseases. Blackwell, Oxford

Crowson RA (1981) The biology of the Coleoptera. Academic Press, London

Curio E (1976) The ethology of predation. Zoophysiology and ecology, vol 7. Springer, Berlin Heidelberg New York

Curtis BA (1988) Do ant-mimicking *Cosmophasis* spiders prey on their *Camponotus* models? Cimbebasia 10:67–70

Dafni A (1984) Mimicry and deception in pollination. Annu Rev Ecol Syst 15:259–278

Dafni A (1986) Floral mimicry-mutualism and undirectional exploitation of insects by flowers. In: Juniper B, Southwood R (eds) Insects and the plant surface. Edward Arnold, London, pp 81–90

Dagg AJ, Foster JB (1976) The giraffe: its biology, behavior and ecology. Van Nostrand Reinhold, New York

Damman H (1986) The osmaterial glands of the swallowtail butterfly *Eurytides marcellus* as a defence against natural enemies. Ecol Entomol 11:261–265

Danin A (1996) Plants of desert dunes (Adaptations of desert organisms). Springer, Berlin Heidelberg New York

Davis J (1973) Habitat preferences and competition of wintering juncos and golden-crowned sparrows. Ecology 54:174–180

Dean WRJ, du Plessis MA, Milton SJ, Adams NJ, Siegfried WR (1992) An association between a stem succulent (*Euphorbia damarana*) and a leaf succulent (*Aloe asperifolia*) in the Namib desert. J Arid Environ 22:67–72

Delany MJ, Happold DCD (1979) Ecology of African mammals. Longman, London

Délye G (1957) Observations sur la fourmi saharienne *Cataglyphis bombycina* Rog. Insectes Soc 4(2):77–82

Dice LR, Blossom PM (1937) Studies of mammalian ecology in south-western North America with special reference to the colors of desert animals. Publ Carnegie Inst Wash 485:1–129

Dobson AP (1989) Behavioral and life history adaptations of parasites for living in desert environments. J Arid Environ 17:185–192

Done BS, Heatwole H (1977) Social behavior of some Australian skinks. Copeia 1977(3):491–430

Donkin RA (1980) Manna: an historical geography. Biogeographica, vol 17. W Junk, The Hague

Dorst J (1974) The life of birds, vol II. (translated by Galbraith ICJ). Weidenfeld & Nicolson, London

Downes MF (1994) Arthropod nest associates of the social spider *Phryganoporus candidus* (Aranea: Desidae). Bull Br Arachnol Soc 9:249–255

Driver PM, Humphries DA (1988) Protean behaviour. The biology of unpredictability. Clarendon Press, Oxford

Duellman WE, Trueb L (1986) Biology of amphibians. McGraw-Hill, New York

Duffey SS (1980) Sequestation of plant natural products by insects. Annu Rev Entomol 25:447–477

Dukas R, Clark CW (1995) Sustained vigilance and animal performance. Anim Behav 49:1259–1267

Dumortier B (1964) Morphology of sound emission apparatus in Arthropoda. In: Busnel R-G (ed) Acoustic behaviour of animals. Elsevier, pp 277–345

Edmunds M (1974) Defence in animals. A survey of antipredator defences. Longman, Harlow

Edwards JS (1961) The action and composition of the saliva of an assassin bug *Platymeris rhadmanthus* Gaerst (Hemiptera: Reduviidae). J Exp Biol 38:61–77

Egoscue HJ (1960) Laboratory and field studies of the northern grasshopper mouse. J Mammal 41:99–110

Ehleringer JR (1993) Variation in leaf carbon isotope discrimination in *Encelia farinosa*: implication for growth, competition and drought survival. Oecologia 95:340–346

Eibl-Eibesfeld I (1979) The biology of peace and war. Men, animals, and aggression (translated by Mosbacher E). Thames and Hudson, London

Eisner T (1962) Survival by acid defense – whip scorpion repels enemies with spray. Nat Hist (NY) 71:10–19

Eisner T, Davis JA (1967) Mongoose throwing and smashing millipedes. Science (NY) 155:577–579

Eisner T, Meinwald J (1966) Defensive secretions of arthropods. Science (NY) 153:1341–1350

Elgar MA (1989) Predator vigilance and group size in mammals and birds: a critical review of the empirical evidence. Biol Rev 64:13–33

Elgar MA (1992) Sexual cannibalism in spiders and other invertebrates. In: Elgar MA, Crespi BJ (eds) Cannibalism. Ecology and evolution among diverse taxa. Oxford University Press, Oxford, pp 128–155

Ellner S, Schmida A (1981) Why are adaptations for long range dispersal rare in desert plants? Oecologia 51:133–144

Eloff FC (1964) On the predatory habits of lions and hyaenas. Koedoe 7:105–112

Enders MM (1995) Size-assortive mating in a tenebrionid beetle of the Namib desert. J Arid Environ 29:469–484

Endler JA (1986) Defense against predators. In: Feder ME, Lauder GV (eds) Predator-prey relationships. Perspectives and approaches from the study of lower vertebrates. University of Chicago Press. Chicago, pp 109–134

Ernest KA (1994) Resistance of creosote bush to mammalian herbivory: temporal consistency and browsing induced changes. Ecology 75:1684–1692

Evans GO, Sheals JG, Macfarlane D (1961) The terrestrial Acari of the British Isles. An introduction to their morphology, biology and classification, vol 1. Introduction and biology. Brit Museum, London

Ewer RF (1968) Ethology of mammals. Logos Press, London

Ewer RF (1973) The carnivores (The World Naturalist). Weidenfeld & Nicolson, London

Fagen R (1981) Animal play behavior. Oxford University Press, New York

Fagerström T (1989) Anti-herbivory chemical defence in plants: a note on the concept of cost. Am Nat 133:281–287

Fenton MB, Licht LE (1990) Why rattle snake? J Herpetol 24:274–279

Fishelson L (1960) The biology and behaviour of *Poekiloceros bufonius* Klug, with a special reference to the repellent gland (Orth Acrididae). Eos Rev Esp Entomol 36:41–62

Fisher RA (1930) The genetical theory of natural selection. Clarendon Press, Oxford

FitzGibbon CD (1990) Mixed-species grouping in Thomson's gazelles: the antipredator benefits. Anim Behav 39:1116–1126

Foelix RF (1982) Biology of spiders. Harvard University Press, Cambridge

Ford HA (1989) Ecology of birds. An Australian perspective. Sutton Beatty, Chipping Norton

Formanowicz DR (1990) The antipredatory efficacy of spider leg autotomy. Anim Behav 40:400–401

Foster JB, Dagg AI (1972) Notes on the giraffe. E Afr Wildl J 10:1–16

Fowler N (1986) The role of competition in plant communities in arid and semiarid regions. Annu Rev Ecol Syst 17:89–110

Fritz RS, Simms EL (eds) (1992) Plant resistance to herbivores and pathogens. Ecology, evolution, and genetics. University of Chicago Press, Chicago

Gans C (1970) How snakes move. Sci Am 222(6):82–96

Gans C (1974) Biomechanics: an approach to vertebrate biology. JP Lippincott, Philadelphia

Garland T Jr (1994) Phylogenetic analyses of lizard endurance capacity in relation to body size and body temperature. In: Vitt LJ, Pianka ER (eds) Lizard ecology. Historical and experimental perspectives. Princeton University Press, Princeton, pp 217–259

Gasaway WC, Mossestad KT, Stander P (1991) Food acquisition by spotted hyaenas in Etosha National Park, Namibia: predation versus scavenging. Afr J Ecol 29:64–75

Gentry AW (1990) Evolution and dispersal of African Bovidae. In: Bubenik GA, Bubenik AB (eds) Horns, pronghorns, and antlers. Evolution, morphology, physiology, and social significance. Springer, Berlin Heidelberg New York, pp 194–227

George U (1969) Über das Tränken der Jungen und andere Lebensäußerungen des Senegal-Flughuhns, *Pterocles senegallus* in Marokko. J Ornithol 110:181–191

George U (1970) Beobachtungen an *Pterocles senegallus* und *Pterocles coronatus* in der Nordwest-Sahara. J Ornithol 111:175–188

Gibson AC (1996) Structure-function relationships of warm desert plants (Adaptations of desert organisms). Springer, Berlin Heidelberg New York

Gier H, Kruckenberg S, Marler R (1978) Parasites and diseases of coyotes. In: Bekoff M (ed) Coyotes: biology, behavior and management. Academic Press, New York, pp 37–69

Gilchrist TL, Stansfield F, Cloudsley-Thompson JL (1966) The odoriferous principle of *Piezodorus teretepes* (Stål) (Hemiptera: Pentatomoidea). Proc R Entomol Soc Lond Ser A Gen Entomol 41:55–56

Gingerich PD (1975) Is the aardwolf a mimic of the hyaena? Nature 253:191–192

Gittleman JL (1989) Carnivore group living: comparative trends. In: Gittleman JL (ed) Carnivore behavior, ecology, and evolution. Cornell University Press, Ithaca; Chapman & Hall, London, pp 183–207

Godfray HCJ (1994) Parasitoids. Behavioral and evolutionary ecology (Monographs in behavior and ecology). Princeton University Press, Princeton

Goodhart CG (1975) Does the aardwolf mimic a hyena? Zool J Linn Soc 57:349–356

Gorvett H (1956) Tegumental glands and terrestrial life in woodlice. Proc Zool Soc Lond 126:291–314

Graetz RD (1981) Plant-animal interactions. In: Goddall DW, Perry RA, Howes KWM (eds) Arid-land ecosystems: structure, function and management (International Biological Programme 17), vol 2. Cambridge University Press, Cambridge, pp 85–103

Grant PR (1972) Interspecific competition among rodents. Annu Rev Ecol Syst 3:79–106

Gray J (1968) Animal locomotion. Weidenfeld & Nicolson, London

Gray R, Bonner J (1948a) An inhibitor of plant growth from the leaves of *Encelia farinosa*. Am J Bot 35:52–57

Gray R, Bonner J (1948b) Structure determination and synthesis of a plant growth inhibitor 3-acetyl-6-methoxy-benzaldehyde found in leaves of *Encelia farinosa*. J Am Chem Soc 70:1249–1253

Greene HW (1988) Antipredator mechanisms in reptiles. In: Gans C, Huey RB (eds) Biology of the Reptilia, vol 16 (Ecology B). Defense and life history. Alan R Liss, New York, pp 1–152

Grenot CJ (1974) Physical and vegetational aspects of the Sahara desert. In: Brown GW Jr (ed) Desert biology. Special topics on the physical and biological aspects of arid regions, vol 2. Academic Press, New York, pp 103–164

Grinnell J, Packer C, Pusey AE (1995) Cooperation in male lions: reciprocity or mutualism? Anim Behav 49:95–105

Gross P (1993) Insect behavioral and morphological defenses against parasitoids. Annu Rev Entomol 38:251–273

Guibé J (1970) Locomotion. In: Grassé PP (ed) Traité de zoologie. Anatomie, systématique, biologie XV Reptiles. Charactères genéraux et anatomie (Fasc 11). Masson, Paris, pp 181–193

Guilford T (1990) The evolution of aposematism. In: Evans DL, Schmidt JO (eds) Insect defenses. Adaptive mechanisms and strategies of prey and predators. State University of New York Press, Albany, pp 23–61

Guilford T (1992) Predator psychology and the evolution of prey coloration. In: Crawley MJ (ed) Natural enemies: the population biology of predators, parasites and diseases. Blackwell, Oxford, pp 377–394

Gulmon SL, Mooney HA (1986) Costs of defense and their effects on plant productivity. In: Givnish TJ (ed) On the economy of plant form and function. Cambridge University Press, Cambridge, pp 681–698

Gutterman Y (1982) Observations on the feeding habits of the Indian crested porcupine (*Hysterix indica*) and the distribution of some hemicryptophytes and geophytes in the Negev desert highlands. J Arid Environ 5:261–268

Gutterman Y (1983) Survival mechanisms of desert winter annual plants in the Negev highlands of Israel. Sci Rev Arid Zone Res 1:249–283

Gutterman Y (1993) Seed germination in desert plants (Adaptations of desert organisms). Springer, Berlin Heidelberg New York

Hadley NF (1972) Desert species and adaptations. Am Sci 60:338–347

Hadley NF (1985) The adaptive role of lipids in biological systems. John Wiley, New York

Hadley NF (1994) Water relations of terrestrial arthropods. Academic Press, San Diego

Hafez M, Makky AMM (1959) Studies on desert insects in Egypt 111. On the bionomics of *Adesmia carinata* Klug. Bull Soc Fouad Entomol 43:89–113

Hall MJR, Smith KGV (1993) Diptera causing myiasis in man. In: Lane PR, Crosskey RW (eds) Medical insects and arachnids. Chapman & Hall, London, pp 429–469

Hamilton WJ III (1973) Life's color code. McGraw Hill, New York

Harborne JB (1993) Introduction to ecological biochemistry, 4th edn. Academic Press, London

Harrison DL (1975) Desert coloration in rodents. In: Prakash I, Ghosh PK (eds) Rodents in desert environments. Dr W Junk, The Hague, pp 269–276

Harvey PH, Gittleman JL (1992) Correlates of carnivory: approaches and answers. In: Crawley MJ (ed) Natural enemies. The population biology of predators, parasites and disease. Blackwell, Oxford, pp 26–30

Hassell MP, Godfray HCJ (1992) The population biology of insect parasitoids. In: Crawley MJ (ed) Natural enemies. The population biology of predators, parasites and diseases. Blackwell, Oxford, pp 265–292

Hayes WK (1995) Venom metering by juvenile prairie rattlesnakes, *Crotalus v. vividis*: effects of prey size and experience. Anim Behav 50:33–40

Heatwole H (1976) Reptile ecology. University Queensland Press, St Lucia

Heatwole H (1996) Energetics of desert invertebrates (Adaptations of desert organisms). Springer, Berlin Heidelberg New York

Heatwole H, Taylor J (1987) Ecology of reptiles. Surrey Beatty, Chipping Norton

Heinrich B (1993) The hot-blooded insects. Strategies and mechanism of thermoregulation. Springer, Berlin Heidelberg New York

Heller J, Ittiel H (1990) Natural history and population dynamics of the land snail *Helix texta* in Israel (Pulmonata: Helicidae). J Molluscan Stud 56:189–204

Henschel JR (1990a) Spiders wheel to escape. S Afr J Sci 86:151–152

Henschel JR (1990b) The biology of *Leucorchestris arenicola* (Araneae: Heteropodidae), a burrowing spider of the Namib dunes. In: Seely MK (ed) Namib ecology: 25 years of Namib research. Transvaal Mus Monogr 7. Transvaal Museum, Pretoria, pp 115–127

Henschel JR (1991) Is solitary life an alternative for the social spider *Stegodyphus dumicola*? J Namib Sci Soc Windhoek 43:71–79

Henschel JR (1994) Diet and foraging behaviour of huntsman spiders in the Namib dunes (Araneae: Heteropodidae). J Zool Lond 234:239–251

Henschel JR, Lubin YD (1992) Environmental factors affecting the web and activity of a psammophilous spider in the Namib desert. J Arid Environ 22:173–189

Henschel JR, Skinner JD (1990) The diet of the spotted hyaenas *Crocuta crocuta* in Kruger National Park. Afr J Ecol 28:69–82

Henschel JR, Mendelsohn JM, Simmons RE (1991) Is the association between the Gabar goshawk and social spiders, *Stegodyphus*, mutualism or theft? Gabar 6:57–60

Herms DA, Mattson WJ (1992) The dilemma of plants: to grow or defend. Q Rev Biol 67:283–335

Herold W (1913) Beiträge zur Anatomie und Physiologie einiger Landisopoden. Zool Jahrb Anat Ontog Tiere 35:457–526

Hertz PE, Huey RB, Nervo E (1982) Fight versus flight: body temperature influences defensive responses of lizards. Anim Behav 30:676–679

Hetherington TE (1989) Use of vibratory cues for detection of insect prey by the sand-swimming lizard *Scincus scincus*. Anim Behav 37:290–297

Hinton HE, Dunn AMS (1967) Mongooses. Their natural history and behaviour. Oliver & Boyd, Edinburgh

Hockey PAR, Boobyer MG (1994) Territoriality and determinants of group size in the Karoo korhaan *Eupodotis vigorsii* (Otididae). J Arid Environ 28:325–332

Hoffmeister DF (1956) Mammals of the Graham (Pinalano) mountains, Arizona. Am Midl Nat 55:257–288

Hofmeyr MD, Louw GN (1987) Thermoregulation, pelage conductance and renal function in the desert-adapted springbok *Antidorcas marsupialis*. J Arid Environ 13:137–151

Hölldobler B, Wilson EO (1990) The ants. Springer, Berlin Heidelberg New York

Holm E, Kirsten JF (1979) Pre-adaptation and speed mimicry among Namib desert scarabaeids with orange elytra. J Arid Environ 2:263–271

Höln R, Petermann J (1980) Curiosities of the plant kingdom (translated by Liebscher H). Cassell, London

Hopkin SP, Read HJ (1992) The biology of millipedes. Oxford University Press, Oxford
Horak IG, deVos D, Braack LEO (1995) Arthropod burdens of impalas in the Skukuza region during two droughts in the Kruger National Park. Koedoe 38(1):65–71
Horner BE, Taylor JM, Padykula HA (1965) Food habits and gastric morphology of the grasshopper mouse. J Mammal 45:513–535
Huey RB, Pianka ER (1977) Natural selection for juvenile lizards mimicking noxious beetles. Science (NY) 195:201–203
Huey RB, Pianka ER (1981) Ecological consequences of foraging mode. Ecology 62:991–999
Huey RB, Pianka ER, Egan ME, Coons LW (1974) Ecological shifts in sympatry: Kalahari fossorial lizards (*Typhlosaurus*). Ecology 55:304–316
Hughes JJ, Ward D (1993) Predation risk and distance covered affect foraging behaviour in Namib desert gerbils. Anim Behav 46:1243–1245
Hughes JJ, Ward D, Perrin MR (1994) Predation risk and competition affect habitat selection and activity of Namib desert gerbils. Ecology 75:1397–1405
Huntingford F, Turner A (1987) Animal conflict (Animal behaviour series). Chapman & Hall, London
Illius AW, FitzGibbon C (1994) Cost of vigilance in foraging ungulates. Anim Behav 47:481–484
Innis AC (1958) The behaviour of the giraffe, *Giraffa camelopardalis*, in the eastern Transvaal. Proc Zool Soc Lond 131:245–278
Isely FB (1938) Survival value of acridian protective coloration. Ecology 19:370–389
Jaeger P (1954) Les aspects actuels du problème de la chéiropterogamie. Bull Inst Fr Afr Noire Sér A Sci Nat 16:786–821
James CD, Hoffman MT, Lightfoot DC, Forbes GS, Whitford WG (1993) Pollination ecology of *Yucca elata*. An experimental study of a mutualistic association. Oecologia 93:512–517
James MT (1947) The flies that cause myiasis in man. US Dept Agric Misc Publ 631:1–175
Jarvis JUM, Bennett NC (1993) Eusociality has evolved independently in two genera of bathyergid mole-rats - but occurs in no other subterranean mammal. Behav Ecol Sociobiol 33:253–260
Jocqué R (1991) A generic revision of the spider family Zodariidae (Araneae). Bull Am Mus Nat Hist 201:1–160
Johnson SD (1994) Evidence for Batesian mimicry in a butterfly-pollinated orchid. Biol J Linn Soc 53:91–104
Jones PJ (1989) Quelea population dynamics. In: Bruggers RL, Elliott CCH (eds) *Quelea quelea*. Africa's bird pest. Oxford University Press, Oxford, pp 198–215
Jones SC, Nutting WL (1989) Foraging ecology of subterranean termites in the Sonoran desert. In: Schmidt JO (ed) Special biotic relationships in the arid southwest. Contributions of the Committee on Desert and Arid Zones Research of the American Association for the Advancement of Science 24. University of New Mexico Press, Albuquerque, pp 80–106
Kachkarov DN, Korovine EP (1942) La vie dans les déserts (Ed française par T Monod). Bibliothèque scientifique. Payot, Paris
Kaestner A (1968) Invertebrate zoology Vol 11. Arthropod relatives, Chelicerata, Mandibulata (translated by Levi HW, Levi LR). Interscience Publishers, New York
Kamil AC, Krebs JR, Pulliam HR (eds) (1987) Foraging behavior. Plenum Press, New York
Karban R (1993a) Induced resistance and plant density of a native shrub *Gossypium thurberi* affect its herbivores. Ecology 74:1–8
Karban R (1993b) Costs and benefits of induced resistance and plant density for a native shrub *Gossypium thurberi*. Ecology 74:9–19
Kerley GIH (1991) Seed removal by rodents, birds and ants in the semi-arid Karoo South Africa. J Arid Environ 20:63–69

Kleiman DG (1977) Monogamy in mammals. Q Rev Biol 52:39–69

Kloos H, McCullough FS (1987) Plants with recognized molluscicidal activity. In: Mott KE (ed) Plant molluscicides: papers presented at a Meeting of the Scientific Working Group on Plant Molluscicides, UNDP/World Bank/WHO Special Programme for Research and Training in Tropical Diseases, held in Geneva, Switzerland, 31 Jan to 2 Feb 1983. John Wiley, Chichester, pp 45–108

Koch C (1955) Monograph of the Tenebrionidae of southern Africa, vol 1. Tentyriinae, Molurini-Trachynotena: *Somaticus* Hope. Transvaal Museum Mem, No 7, Pretoria

Kotler BP (1984) Risk of predation and the structure of desert rodent communities. Ecology 65:689–701

Kotler BP, Brown JS (1988) Environmental heterogeneity and the coexistence of desert rodents. Annu Rev Ecol Syst 19:281–307

Kotler BP, Holt RD (1989) Predation and competition: the interaction of two types of species interactions. Oikos 54:256–260

Kotler BP, Brown JS, Smith RJ, Wirtz WO II (1988) The effects of monophagy and body size on rates of owl predation on desert rodents. Oikos 53:145–152

Kotler BP, Brown JS, Slotow RH, Goodfriend WL, Strauss M (1993) The influence of snakes on the foraging behavior of gerbils. Oikos 67:309–316

Krapf D (1986) Predator prey relations in diurnal *Scorpio maurus*. L Actas X Congr Int Arachnol (Jaca, Spain)1:133

Krebs CJ (1985) Ecology. The experimental analysis of distribution and abundance, 3rd edn. Harper & Row, New York

Krivokhatskiy VA (1985) On the history of formation of the nidicolous entomofauna of sandy deserts of central Asia. Entomol Obozren 4:696–704

Kroll JC (1977) Self-wounding while death feigning by western hognose snakes (*Heterodon nasicus*). Copeia(2):372–373

Kruuk H (1966) Clan-system and feeding habits of spotted hyaenas (*Crocuta crocuta* Erxleben). Nature 209:1257–1258

Kruuk H (1967) Competition for food between vultures in East Africa. Ardea 55:172–193

Kruuk H (1972) The spotted hyaena. A study of predation and social behaviour. University of Chicago Press, Chicago

Kruuk H (1975a) Functional aspects of social hunting by carnivores. In: Baerends G, Beer C, Manning A (eds) Function and evolution in behaviour. Clarendon Press, Oxford, pp 119–141

Kruuk H (1975b) Hyaena. Oxford University Press, London

Kruuk H (1976) Feeding and social behaviour of the striped hyaena (*Hyaena vulgaris* Desmarest). E Afr Wildl J 14:91–111

Kruuk H, Macdonald DW (1985) Group territories of carnivores: empires and enclaves. In: Sibly R, Smith R (eds) Ecological consequences of adaptive behaviour. Blackwell, Oxford, pp 521–536

Kruuk H, Mills MGL (1983) Notes on food and foraging of the honey badger *Mellivora capensis* in the Kalahari Gemsbok National Park. Koedoe 26:153–157

Kruuk H, Sands WA (1972) The aardwolf (*Proteles cristatus*) Sparrman 1783 as predator of termites. E Afr Wildl J 10:211–227

Kruuk H, Turner M (1967) Comparative notes on predation by lion, leopard, cheetah and wild dog in the Serengeti area, East Africa. Mammalia 31:1–27

Lane C, Rothschild M (1965) A case of Müllerian mimicry of sound. Proc R Ento Soc Lond Ser A Gen Entomol 40:156–158

Langley W (1981) The effect of prey defenses on the attack behavior of the southern grasshopper mouse (*Onychomys torridus*). Z Tierpsychol 56:115–127

Lawrence RF (1984) The centipedes and millipedes of southern Africa. A guide. AA Balkema, Cape Town

Le Berre M (1979) Analyse sequentielle du comportement alimentaire du scorpion *Buthus occitanus* (Amor.) (Arach. Scorp. Buth.). Biol Behav 4:97–122

Le Berre M (1989) Faune du Sahara 1 Poissons-amphibiens-reptiles. Lechevalier-R Chabaud, Paris

Lederhouse RC (1990) Avoiding the hunt: primary defenses of lepidopteran caterpillars. In: Evans DL, Schmidt JO (eds) Insect defenses. Adaptive mechanisms and strategies of prey and predators. State University of New York Press, Albany, pp 175–189

Lee PLM, Clayton DH (1995) Population biology of swift (*Apus apus*) ectoparasites in relation to host reproductive success. Ecol Entomol 20:43–50

Lehmann T (1993) Ectoparasitism: desert impact on host fitness. Parasitol Today 9:8–13

Leuthold W (1977) African ungulates. A comparative review of their ethology and behavioral ecology. Springer, Berlin Heidelberg New York

Lewis JGE (1964) Protective devices in two species of African Coleoptera. Proc R Ent Soc Lond Ser A Gen Entomol 39:50–52

Lewis JGE (1981) The biology of centipedes. Cambridge University Press, Cambridge

Lima SL (1994) On the personal benefits of anti-predatory vigilance. Anim Behav 48:734–736

Lima SL (1995) Back to the basics of anti-predatory vigilance: the group-size effect. Anim Behav 49:11–20

Linsenmair KE (1984) Comparative studies on the social behaviour of the desert isopod *Hemilepistus reaumuri* and of a *Porcellio* species. Symp Zool Soc Lond 53:423–453

Little RJ (1983) A review of floral food deception mimicries with comments on floral mutualism. In: Jones CE, Little RJ (eds) Handbook of experimental pollination biology. Scientific and Academic Editions, New York, pp 294–309

Little RM, Earlé RA (1995) Sandgrouse (Pterocleidae) and sociable weavers (*Philetarius socius*) lack avian Haematozoa in semi-arid regions of South Africa. J Arid Environ 30:367–370

Londt JGH (1991) *Bana*, a new genus of bee-mimicking assassin fly from southern Namibia (Diptera: Asilidae: Stenopogoninae). Cimbebasia 13:91–97

Lourenço WR, Cloudsley-Thompson JL (1995) Stridulatory apparatus and the evolutionary significance of sound production in *Rhopalurus* species (Scorpiones: Buthidae). J Arid Environ 31:423–429

Louw GN (1972) The role of advective fog in the water economy of certain Namib desert animals. In: Maloiy GMO (ed) Comparative physiology of desert animals. Symp Zool Soc Lond No 31:297–314

Louw GN, Seely MK (1982) Ecology of desert organisms. Longman, London

Lovegrove B (1993) The living deserts of southern Africa. Fernwood Press, Newlands

Lubin YD (1986) Web building and prey capture in the Uloboridae. In: Shear WA (ed) Spiders. Webs, behavior and evolution. Stanford University Press, Stanford, pp 132–171

Lubin YD, Henschel JR (1990) Foraging at the thermal limit: burrowing spiders (*Seothyra*, Eresidae) in the Namib desert dunes. Oecologia 84:461–467

Ludwig JA (1986) Primary production variability in desert ecosystems. In: Whitford WG (ed) Pattern and process in desert ecosystems. Contributions of the Committee on Desert and Arid Zones Research of the Southeastern and Rocky Mountain division of the American Association for the advancement of Science 21. University of New Mexico Press, Albuquerque, pp 5–17

Ludwig JA, Whitford WG (1981) Short-term water and energy flow in arid ecosystems. In: Goodall DW, Perry RA, Howes KMW (eds) Arid-land ecosystems: structure, functioning and management (International biological programme 17), vol 2. Cambridge University Press, London, pp 271–299

Luke C (1986) Convergent evolution of lizard toe fringes. Biol J Linn Soc 27:1–16

MacKay WP (1991) The role of ants and termites in desert communities. In: Polis GA (ed) The ecology of desert communities. University of Arizona Press, Tucson, pp 113–150

Maclean GL (1968) Field studies on the sand grouse of the Kalahari Desert. Living Bird 7:209–235

Maclean GL (1976) Adaptations of sandgrouse for life in arid lands. Proc 16 Int Ornithol Congr Canberra 1974, pp 502–516

Maclean GL (1996) Ecophysiology of desert birds (Adaptations of desert organisms). Springer, Berlin Heidelberg New York

Madden D, Young TP (1992) Symbiotic ants as an alternative defence against giraffe herbivory in spinescent *Acacia drepanolobium*. Oecologia 91:235–238

Main BY (1976) Spiders (The Australian naturalist library). Collins, Sydney

Marden JH (1987) In pursuit of females: following and contest behaviour by males of a Namib desert tenebrionid beetle *Physadesmia globosa*. Ethology 75:15–24

Martins EP (1994) Phylogenetic perspectives on the evolution of lizard territoriality. In: Vitt LJ, Pianka ER (eds) Lizard ecology. Historical and experimental perspectives. Princeton University Press, Princeton, pp 117–144

Mayhew WW (1968) Biology of desert amphibians and reptiles. In: Brown GW Jr (ed) Desert biology. Special topics on the physical and biological aspects of arid regions, vol 1. Academic Press, London, pp 195–356

McAuliffe JR (1984) Sahuaro-nurse tree associations in the Sonoran desert: competitive effects of sahuaros. Oecologia 64:319–321

McCleary JA (1968) The biology of desert plants. In: Brown GW Jr (ed) Desert biology. Special topics on the physical and biological aspects of arid regions, vol 1. Academic Press, New York, pp 141–194

McCormick SJ, Polis GA (1990) Prey, predators and parasites. In: Polis GA (ed) The biology of scorpions. Stanford University Press, Stanford, pp 294–320

McIver JD, Stonedahl G (1993) Myrmecomorphy: morphological and behavioral mimicry of ants. Annu Rev Entomol 38:351–379

McLachlan A (1991) Ecology of coastal dune fauna. J Arid Environ 21:229–243

McNaughton SJ, Georgiadis NJ (1986) Ecology of African grazing and browsing mammals. Annu Rev Ecol Syst 17:39–65

Medel RG (1995) Convergence and historical effects in harvester ant assemblages of Australia, North America, and South America. Biol J Linn Soc 55:29–44

Meikle Griswold T (1986) Nest associates of two species of group-living Eresidae in southern Africa. Actas X Congr Int Arachnol (Jaca, Spain)1:275

Mendelssohn H (1963) On the biology of the venomous snakes of Israel, Part 1. Isr J Zool 12:143–170

Menzel R, Shmida A (1993) The ecology of flower colours and the natural colour vision of insect pollinators: the Israeli flora as a study case. Biol Rev 68:81–120

Meserve PL, Gautierrez JR, Jaksic FM (1993) Effects of vertebrate predation on a caviomorph rodent, the degu (*Octodon degus*), in a semiarid thorn scrub community in Chile. Oecologia 94:153–158

Milewski AV, Truman PY, Madden D (1991) Thorns as induced defenses: experimental evidence. Oecologia 86:70–75

Miller MF (1994) Large African herbivores, bruchid beetles and their interactions with *Acacia* seeds. Oecologia 97:265–270

Mills MGL (1990) Kalahari hyaenas. The comparative behavioral ecology of two species. Chapman & Hall, London

Mloszewski MJ (1983) The behavior and ecology of the African buffalo. Cambridge University Press, Cambridge

Moller AP (1990) Effects of parasitism by the haematophagous mite *Ornithonyssus bursa* on reproduction in the barn swallow *Hirundo rustica*. Ecology 71:2345–2357

Monod T (1954) Modes "contracté" et "diffus" de la végétation saharienne. In: Cloudsley-Thompson JL (ed) Biology of deserts. Proc Symp on the Biology of hot and cold deserts, organized by the Institute of Biology. Institute of Biology, London, pp 35–44

Moore BP (1974) The larval habits of two species of *Sphallomorpha* Westwood (Coleoptera: Carabidae, Pseudomorphinae). J Austr Entomol Soc 13:179–183

Mooring MS (1995) The effect of tick challenge on the grooming rate by impala. Anim Behav 50:377–392

Morton SR (1979) Diversity of desert-dwelling mammals: a comparison of Australia and North America. J Mammal 60:253–264

Morton SR, Davidson DW (1988) Comparative structure of harvester ant communities in arid Australia and North America. Ecol Monogr 58:19–38

Mosauer W (1932) On the locomotion of snakes. Science (NY) 76:583–585

Mott JJ (1979) Flowering, seed formation and dispersal. In: Goodall DW, Perry RA, Howes KMW (eds) Arid-land ecosystems: structure, functioning and management (International Biological Programme 16) vol 1. Cambridge University Press, Cambridge, pp 627–645

Müller F (1879) *Ituma* and *Thyridia*; a remarkable case of mimicry in butterflies. (translated by Mendola R) Proc Entomol Soc Lond XX–XXIX

Murray KG, Russell S, Picone CM, Winnett-Murray K, Sherwood W, Kuhlmann M (1994) Fruit laxatives and seed passage rates in frugivores: consequences for plant reproductive success. Ecology 75:989–994

Nelson B (1973) Azraq: desert oasis. Lane, London

Nentwig W (1987) The prey of spiders. In: Netwig W (ed) Ecophysiology of spiders. Springer, Berlin Heidelberg New York, pp 249–263

Newman RA (1994) Effects of changing density and food level on metamorphosis of a desert amphibian, *Scaphiopus couchii*. Ecology 75:1085–1096

Nicolai Y (1989) Thermal properties and fauna on the bark of trees in two different African ecosystems. Oecologia 80:421–430

Niethammer G (1959) Die Rolle der Auslese durch Feinde bei Wüstenvögeln. Bonn Zool Beitr 10:179–197

Niklas KJ (1992) Plant biomechanics. An engineering approach to plant form and function. University Chicago Press, Chicago

Noble JC (1975) The effects of emus (*Dromaius novaehollandiae* Latham) on the distribution of the nitre bush (*Nitraria billardieri* DC). J Ecol 63:979–984

Noy-Meir I (1973) Desert ecosystems: environment and producers. Annu Rev Ecol Syst 4:25–51

Orians G (1971) Ecological aspects of behaviour. In: Farner DS, King JR, Parkes KC (eds) Avian biology, vol 1. Academic Press, New York, pp 513–546

Owen DF (1990) The language of attack and defence. Oikos 57:133–135

Owen J (1980) Feeding strategy (Survival in the wild). Oxford University Press, Oxford

Paarmann W (1979) A reduced number of larval instars, as an adaptation of the desert carabid beetle *Thermophilum (Anthia) sexmaculatum* F. (Coleoptera, Carabidae) to its arid environment. Misc Pap 18:113–117

Paarmann W (1985) Larvae preying on ant broods: an adaptation of the desert carabid beetle *Graphipterus serrator* Forskäl (Col., Carabidae) to arid environments. J Arid Environ 9:210–214

Packer C, Pusey AE, Rowley H, Gilbert DA, Martenson J, O'Brien SJ (1991) Case study of a population bottleneck: lions of the Ngorongoro crater. Conserv Biol 5:219–230

Pantastico-Caldas M, Venable DL (1993) Competition in two species of desert annuals along a topographic gradient. Ecology 74:2192–2203

Parker MA, Root RB (1981) Insect herbivores limit habitat distribution of a native composite *Machaeranthera canescens*. Ecology 62:1390–1392

Parker WS, Pianka ER (1974) Further ecological observations on the western banded gecko, *Coleonyx variegatus*. Copeia 1974(4):615–632

Peckham EG (1889) Protective resemblances in spiders. Occ Pap Nat Hist Soc Wisc 1:61–113

Perrin MR, Boyer DC (1994) The effect of rodents on plant recruitment and production in the dune fields of the Namib desert. Trop Zool 7:299–308

Peters HN (1993) On the burrowing behaviour and the production and use of silk in *Seothyra*, a sand-inhabiting spider from the Namib desert (Arachnida, Araneae, Eresidae). Verh Naturwiss Ver Hamb (NF) 33(1992):191–211

Pettey FW (1947) The biological control of prickly pears in South Africa. Union S Afr Dept Agric Sci Bull 271:1–161

Pianka ER (1972) *r* and K selection or *b* and *d* selection? Am Nat 106:581–588

Pianka ER (1973) The structure of lizard communities. Annu Rev Ecol Syst 4:53–74

Pianka ER (1975) Niche relations of desert lizards. In: Cody ML, Diamond JM (eds) Ecology and evolution of communities. Harvard University Press (Belknap Press), Cambridge, pp 292–314

Pianka ER (1978) Evolutionary ecology, 2nd edn. Harper & Row, New York

Pianka ER (1985) Some intercontinental comparisons of desert lizards. Natl Geogr Res 1:490–504

Pianka ER (1986) Ecology and natural history of desert lizards. Analysis of the ecological niche and community structure. Princeton University Press, Princeton

Pierre F (1958) Écologie et peuplement entomologique des sables vifs du Sahara nord-occidental. Centre National de la Recherche Scientifique, Paris

Poinar G (1985) Mermethid (Nematoda) parasites of spiders and harvestmen. J Arachnol 13:121–128

Polis GA (1979) Prey and feeding phenology of the desert sand scorpion *Paruroctonus mesaensis* (Scorpionidae: Vaejovidae). J Zool Lond 188:333–346

Polis GA (1980) The effect of cannibalism on the demography and activity of a natural population of desert scorpions. Behav Ecol Sociobiol 7:25–35

Polis GA (1981) The evolution and dynamics of intraspecific predation. Annu Rev Ecol Syst 12:225–251

Polis GA (ed) (1990a) The biology of scorpions. Stanford University Press, Stanford

Polis GA (1990b) Ecology. In: Polis GA (ed) The biology of scorpions. Stanford University Press, Stanford, pp 247–293

Polis GA (ed) (1991a) The ecology of desert communities. University of Arizona Press, Tucson

Polis GA (1991b) Food webs in desert communities: complexity via diversity and omnivory. In: Polis GA (ed) (1991a) The ecology of desert communities. University of Arizona Press, Tucson, pp 383–437

Polis GA (1991c) Complex trophic interaction in deserts: an empirical critique of food-web theory. Am Nat 138:123–155

Polis GA, McCormick SJ (1986) Scorpions, spiders and solpugids, predation and competition among distantly related taxa. Oecologia 71:111–116

Polis GA, Myers CA, Holt RD (1989) The ecology and evolution of intraguild predation: potential competitors that eat each other. Annu Rev Ecol Syst 20:297–330

Pough RH (1969) The morphology of undersand respiration in reptiles. Herpetologia 25:216–223

Pough RH (1970) The burrowing ecology of the sand lizard *Uma notata*. Copeia 1970(1):145–157

Proctor M, Yeo P (1973) The pollination of flowers (The New Naturalist). Collins, London

Rand AS (1994) Behavioral ecology. In: Vitt LJ, Pianka ER (eds) Lizard ecology. Historical and experimental perspectives. Princeton University Press, Princeton, pp 91–93

Rasa AE (1987) Sociability for survival: why dwarf mongooses live in groups. In: Drickamer LC (ed) Readings from the 19th Int Ethology Conf, vol 4. Behavioural ecology and population biology. Privat Press, Toulouse, pp 35–39

Rasa AE (1989) Helping in dwarf mongoose societies: an alternative reproductive strategy. In: Rasa AE, Vogel C, Voland E (eds) The sociobiology of sexual and reproductive strategies. Chapman & Hall, London, pp 61–73

Regal PJ (1978) Behavioral differences between reptiles and mammals: an analysis of activity and mental capabilities. In: Greenberg N, MacLean PD (eds) Behavior and neurology of lizards. National Institute of Health, Washington, DC, pp 183–202

Rhoades DF (1979) Evolution of plant chemical defense against herbivores. In: Rosenthal GA, Janzen DH (eds) Herbivores: their interaction with secondary plant metabolites. Academic Press, New York, pp 3–54

Ribble DO, Salvioni M (1990) Social organization and nest co-occupancy in *Peromyscus californicus*, a monogamous rodent. Behav Ecol Sociobiol 26:9–15

Rice EL (1984) Allelopathy, 2nd edn. Academic Press, Orlando

Richards OW, Davies RG (1977) Imms' general textbook of entomology, 10th edn. (2 vols). Chapman & Hall, London. John Wiley, New York

Ridley HN (1930) The dispersal of plants throughout the world. L Reeve, Ashford

Riechert SE (1979) Development and reproduction. In: Goodall DW, Perry RA, Howes KMW (eds) Arid-land ecosystems: structure, functioning or management (International biological programme 16), vol 1. Cambridge University Press, Cambridge, pp 797–822

Rood JP (1983) The social system of the dwarf mongoose. In: Eisenberg JF, Kleiman DG (eds) Recent advances in the study of mammalian behavior. American Society of Mammalogists, Shippensburg, pp 454–458

Rood JP (1990) Group size, survival, reproduction and routes to breeding in dwarf mongooses. Anim Behav 39:566–572

Root TM (1990) Neurobiology. In: Polis GA (ed) The biology of scorpions. Stanford University Press, Stanford, pp 341–413

Rosenthal GA, Berenbaum MR (eds) (1991) Herbivores. Their interactions with secondary plant metabolites, vol 1, 2nd edn. The chemical participants. Academic Press, San Diego

Rosenthal GA, Berenbourm MR (eds) (1992) Herbivores. Their interactions with secondary plant metabolites, vol II, 2nd edn. Ecological and evolutionary processes. Academic Press, San Diego

Rothschild M, Clay T (1952) Fleas, flukes and cuckoos. A study of bird parasites (The New Naturalist). Collins, London

Rydell J, Speakman JR (1995) Evolution of nocturnality in bats: potential competitors and predators during their early history. Biol J Linn Soc 54:183–191

Samson DA, Philippi TE, Davidson DW (1992) Granivory and competition as determinants of annual plant diversity in the Chihuahuan desert. Oikos 65:61–80

Savitzky AH (1981) Hinged teeth in snakes: an adaptation for swallowing hard-bodied prey. Science (NY) 212:346–349

Schall JJ, Marghoob AB (1995) Prevalence of a malarial parasite over time and space: *Plasmodium mexicanum* in its vertebrate host, the western fence lizard *Sceloporus occidentalis*. J Anim Ecol 64:177–185

Schall JJ, Pianka ER (1978) Geographical trends in numbers of species. Science (NY) 201:679–686

Schaller GB (1972) The Serengeti lion. A study of predator-prey relations. University of Chicago Press, Chicago

Schmidt JO (1989) Spines and venoms in flora and fauna in the Arizona upland Sonoran desert and how they act as defenses against vertebrates. In: Schmidt JO (ed) Special biotic relationships in the arid southwest. Contributions of the Committee on Desert and Arid Zones Research of the Southwestern and Rocky Mountain Division of the American Association for the Advancement of Science 24. Univ New Mexico Press, Albuquerque, pp 107–129

Schmidt JO (ed) (1990) Special biotic relationships in the arid southwest. Contributions of the Committee on Desert and Arid Zones Research of the Southwestern and Rocky Mountain Division of the American Association for the Advancement of Science 24. University of New Mexico Press, Albuquerque

Schmidt PJ, Schmidt JO (1989) Harvester ants and horned lizards. Predator-prey interactions. In: Schmidt JO (ed) Special biotic relationships in the arid southwest. Contributions of the Committee on Desert and Arid Zones Research of the Southwestern and Rocky Mountain Division of the American Association for the Advancement of Science 24. University New Mexico Press, Albuquerque, pp 25–51

Schoener TW (1971) Theory of feeding strategies. Annu Rev Ecol Syst 2:369–404

Schwarzkopf L (1994) Measuring trade-offs: a review of studies of costs of reproduction in lizards. In: Vitt LJ, Pianka ER (eds) Lizard ecology. Historical and experimental perspectives. Princeton University Press, Princeton, pp 7–29

Schwarzkopf L, Shine R (1992) Costs of reproduction in lizards: escape tactics and susceptibility to predation. Behav Ecol Sociobiol 31:17–25

Seely MK (1978) The Namib dune desert: an unusual ecosystem. J Arid Environ 1:117–128

Seely MK (1987) The Namib. Natural history of an ancient desert. Shell Oil SWA, Windhoek

Seely MK, Louw GN (1980) First approximation of the effects of rainfall on the ecology and energetics of rainfall on the ecology and energetics of a Namib dune ecosystem. J Arid Environ 3:25–54

Serventy DL (1971) Biology of desert birds. In: Farner SD, King JR, Parkes KC (eds) Avian biology, vol 1. Academic Press, New York, pp 287–339

Shachak M, Brand S (1983) The relationship between sit and wait foraging strategy and dispersal in the desert scorpion, *Scorpio maurus palmatus*. Oecologia 60:371–377

Sharman GB, Calaby JH (1964) Reproductive behaviour in the red kangaroo, *Megaleia rufa*, in captivity. CSIRO Wildl Res 9:58–85

Sherbrooke WC, Montanucci RR (1988) Stone mimicry in the round-tailed horned lizard, *Phrynosoma modestum* (Sauria: Iguanidae). J Arid Environ 14:275–284

Sherman PW, Jarvis JVM, Alexander RD (eds) (1991) The biology of the naked mole rat. Princeton University Press, Princeton

Shields WM (1984) Barn swallow mobbing: self-defence, collateral kin defence, group defence, or parental care? Anim Behav 32:132–148

Shine R (1980) "Costs" of reproduction in reptiles. Oecologia 46:92–100

Shine R (1988) Parental care in reptiles. In: Gans C, Huey RB (eds) Biology of the Reptilia 16 (Ecology B). Defense and life history. Alan R Liss, New York, pp 275–329

Shreve F (1910) The rate of establishment of the giant cactus. Plant World 13:235–240

Sih A (1993) Effects of ecological interactions on forager diets: competition, predation risk, parasitism and prey behaviour. In: Hughes RN (ed) Diet selection. An interdisciplinary approach to foraging behaviour. Blackwell, Oxford, pp 182–211

Silvertown J, Wilson JB (1994) Community structure in a desert perennial community. Ecology 75:409–417

Simard JM, Watt DD (1990) Venoms and toxins. In: Polis GA (ed) The biology of scorpions. Stanford University Press, Stanford, pp 414–444

Simmons LW (1990) Post-copulatory guarding, female choice and the levels of gregarine infections in the field cricket *Gryllus bimaculatus*. Behav Ecol Sociobiol 26:403–407

Sinclair ARE (1977) The African buffalo. A study of resource limitation of populations (Wildlife behavior and ecology). University of Chicago Press, Chicago

Sinervo B, Hedges R, Adolph SC (1991) Decreased sprint speed as a cost of reproduction in the lizard *Sceloporus occidentalis*: variation among populations. J Exp Biol 155:323–336

Skogsmyr I, Fagerström T (1992) The cost of anti-herbivory defence: an evaluation of some ecological and physiological factors. Oikos 64:451–457

Slotow R, Goodfriend W, Ward D (1993) Shell colour polymorphism of the Negev desert land snail, *Trochoidea sectzeni*: the importance of temperature and predation. J Arid Environ 24:47–61

Smith KGV (1993) Insects of minor medical importance. In: Lane RP, Crosskey RW (eds) Medical insects and arachnids. Chapman & Hall, London, pp 576–593

Smith SD, Monson RK, Anderson TE (1996) Physiological ecology of North American desert plants (Adaptations of desert organisms). Springer, Berlin Heidelberg New York

Soholt LT (1973) Consumption of primary production by a population of kangaroo rats (*Dipodomys merriami*) in the Mojave desert. Ecol Monogr 43:357–376

Sømme L (1995) Invertebrates in hot and cold arid environments (Adaptations of desert organisms). Springer, Berlin Heidelberg New York

Sorci G, Massot M, Clobert J (1994) Maternal parasite load increases sprint speed and philopatry in female offspring of the common lizard. Am Nat 144:153–164

Sorensen AE (1986) Seed dispersal by adhesion. Annu Rev Ecol Syst 17:443–463

Speed MP (1993) When is mimicry good for predators? Anim Behav 46:1246–1248

Spinage CA (1986) The natural history of antelopes (Croom Helm mammal series). Croom Helm, London

Spradbery JP (1973) Wasps. An account of the biology and natural history of solitary and social wasps. University of Washington Press, Seattle

Springett BP (1968) Aspects of the relationships between burying beetles, *Necrophorus* spp. and the mite *Poecilochirus necrophori* Vitz. J Anim Ecol 37:417–424

Stamps JA (1983) Sexual selection, sexual dimorphism and territoriality. In: Huey RB, Pianka R, Schoener TW (eds) Lizard ecology. Studies of a model organisms. Harvard University Press, Cambridge, pp 169–204

Stamps JA, Krishnan VV (1994a) Territory acquisition in lizards: I. First encounters. Anim Behav 47:1375–1385

Stamps JA, Krishnan VV (1994b) Territory acquisition in lizards: II. Establishing social and spatial relationships. Anim Behav 47:1387–1400

Stamps JA, Krishnan VV (1995) Territory acquisition in lizards: III. Competing for space. Anim Behav 49:679–693

Starrett A (1993) Adaptive resemblance: a unifying concept for mimicry and crypsis. Biol J Linn Soc 28:299–317

Stearns SC (1976) Life-history tactics: a review of the ideas. Q Rev Biol 51:3–47

Stebbins RC (1943) Adaptations of the nasal passages for sand burrowing in the saurian genus *Uma*. Am Nat 77:38–52

Steenbergh WF, Lowe CH (1977) Ecology of the saguaro: 11. Reproduction, germination, establishment, growth and survival of the young plant. National Parks Service Sci Monogr Ser No 8, Washington, DC

Stevens L (1992) Cannibalism in beetles. In: Elgar MA, Crespi BJ (eds) Cannibalism. Ecology and evolution among diverse taxa. Oxford University Press, Oxford, pp 156–175

Sweet SS (1985) Geographic variation, convergent crypsis and mimicry in gopher snakes (*Pituophis melanoleucus*) and western rattlesnakes (*Crotalus viridis*). J Herpetol 19:55–67

Taylor ME (1989) Locomotor adaptation by carnivores. In: Gittleman JL (ed) Carnivore behavior, ecology, and evolution. Cornell University Press, Ithaca; Chapman & Hall, London, pp 382–409

Telford S (1971) A comparative study of endoparasitism among some California lizard populations. Am Midl Nat 83:516–554

Terrick TD, Mumme RL, Burghardt GM (1995) Aposematic coloration enhances chemosensory recognition of noxious prey in the garter snake *Thamnophis radix*. Anim Behav 49:857–866

Tevis L Jr, Newell IM (1962) Studies on the biology and seasonal cycle of the giant red velvet mite *Dinothrombium pandorae* (Acari, Trombidiidae). Ecology 43:497–505

Thiollay J-M (1989) Natural predation on quelea. In: Bruggers RL, Elliott CCH (eds) *Quelea quelea* Africa's bird pest. Oxford University Press, Oxford, pp 216–229

Thorpe WH (1942) Observations on *Stomoxys ochrosoma* (Diptera: Muscidae) as an associate of army ants (Dorylinae) in East Africa. Proc R Ent Soc Lond Ser A Gen Entomol 17:38–41

Tietjen WJ (1986) Social spider webs, with special reference to the web of *Mallos gregalis*. In: Shear WA (ed) Spiders. Webs, behavior and evolution. Stanford University Press, Stanford, pp 172–206

Tinsley RC (1989) Effects of host sex on transmission success. Parasitol Today 5:190–195

Tinsley RC (1990) The influence of parasite infection on mating success in spadefoot toads, *Scaphiopus couchii*. Am Zool 30:313–324

Trägårdh I (1943) Die Milben und ihre ökologischen Beziehungen zu den Insekten. Arb Physiol Angew Entomol Berl 10:124–136

Urquhart FA (1960) The monarch butterfly. University of Toronto Press, Toronto

Valone TJ, Brown JH, Heske EJ (1994) Interactions between rodents and ants in the Chihuahuan desert: an update. Ecology 75:252–255

Van Lawick-Goodall J (1968) Tool-using bird: the Egyptian vulture. Natl Geogr Mag 133:630–641

Van Valkenburgh B (1989) Carnivore dental adaptation and diet: a study of trophic diversity within guilds. In: Gittleman JL (ed) Carnivore behavior, ecology and dentition. Cornell University Press, Ithaca; Chapman & Hall, London, pp 410–436

Van Valkenburgh B, Wayne RK (1994) Shape divergence associated with size convergence in sympatric East African jackals. Ecology 75:1567–1581

Vane-Wright RI (1980) On the definition of mimicry. Biol J Linn Soc 13:1–6

Vane-Wright RI (1981) Only connect. Biol J Linn Soc 16:33–40

Vetter RS (1980) Defensive behavior of the black widow spider *Latrodectus hesperus* (Araneae: Theridiidae). Behav Ecol Sociobiol 7:187–193

Waage J, Greathead P (eds) (1986) Insect parasitoids (13th Symp, Royal Entomological Society of London). Academic Press, London

Wagner FH, Graetz RD (1981) Animal-animal interactions. In: Goodall DW, Perry RA, Howes KMW (eds) Arid-land ecosystems: structure, functioning and management (International Biological Programme 17), vol 2. Cambridge University Press, Cambridge, pp 51–83

Wallwork JA (1982) Desert soil fauna. Praeger, New York

Walter H, Stadelmann E (1974) A new approach to the water relations of desert plants. In: Brown GW Jr (ed) Desert biology, vol II. Academic Press, New York, pp 213–310

Walther FR (1969) Flight behaviour and avoidance of predators in Thomson's gazelle (*Gazella thomsoni* Guenther 1884). Behaviour 34:184–221

Warburg MR (1993) Evolutionary biology of land isopods. Springer, Berlin Heidelberg New York

Ward D, Henschel JR (1992) Experimental evidence that a desert parasitoid keeps its host cool. Ethology 92:135–142

Ward D, Saltz D (1994) Foraging at differential spatial scales: dorcas gazelles foraging for lilies in the Negev desert. Ecology 75:48–58

Ward P (1979) Colour for survival. Orbis, London

Warrag MOA (1994) Autotoxicity of mesquite (*Prosopis juliflora*) pericarp on seed germination and seedling growth. J Arid Environ 27:79–84

Watson RT, Irish J (1988) An introduction to the Lepismatidae (Thysanura: Insecta) of the Namib desert sand dunes. Madoqua 15:285–293

Wharton RA (1987) Biology of the diurnal *Metasolpuga picta* (Kraepelin) (Solifugae, Solpugidae) compared with that of nocturnal species. J Arachnol 14:363–383

Wheeler WM (1930) Demons of the dust. WW Norton, New York

Whitford WG (ed) (1986a) Pattern and process in desert ecosystems. Contributions of the Committee on Desert and Arid Zones Research of the Southwestern and Rocky Mountain Division of the American Association for the Advancement of Science 21. University New Mexico Press, Albuquerque

Whitford WG (1986b) Decomposition and nutrient cycling in deserts. In: Whitford WG (ed) Pattern and process in desert ecosystems. Contributions of the Committee on Desert and Arid Zones Research of the Southwestern and Rocky Mountain Division of the American Association for the Advancement of Science 21. University of New Mexico Press, Albuquerque, pp 93–117

Whitman DW, Blum MS, Jones CG (1986) Prey-specific attack behaviour in the southern grasshopper mouse *Onychomys torridus* (Coues). Anim Behav 34:295–297

Whitman DW, Blum MS, Alsop DW (1990) Allomones: chemicals for defence. In: Evans DL, Schmidt JO (eds) Insect defenses. Adaptive mechanisms and strategies of prey and predators. State University of New York Press, Albany, pp 289–351

Wickler W (1968) Mimicry in plants and animals (translated by Martin RD). Weidenfeld & Nicolson, London. McGraw-Hill, New York

Wiens D (1982) Mimicry in plants. Evol Biol 11:365–403

Wilczynski W (1992) The nervous system. In: Feder ME, Burggren WW (eds) Environmental physiology of the amphibians. University of Chicago Press, Chicago, pp 9–39

Williams SC (1987) Scorpions bionomics. Annu Rev Entomol 32:275–295

Willis EO (1963) Is the zone-tailed hawk a mimic of the turkey vulture? Condor 65:313–317

Willoughby EJ (1969) Desert coloration in birds of the central Namib desert. Sci Pap Namib Desert Res St 44:59–68

Wilson EO (1971) The insect societes. Belknap Press of Harvard University Press, Cambridge

Wilson RT (1984) The camel. Longman, London

Winemiller KO, Pianka ER (1990) Organization in natural assemblages of desert lizards and tropical fishes. Ecol Monogr 60:27–55

Winkler DW (1994) Anti-predator defence by neighbours as a responsive amplifier of parental defence in tree swallows. Anim Behav 47:595–605

Wittenberger JF, Hunt GL Jr (1985) The adaptive significance of coloniality in birds. In: Farner DS, King JR, Parkes KC (eds) Avian biology, vol VIII. Academic Press, Orlando, pp 1–78

Wittenberger JF, Tilson RL (1980) The evolution of monogamy: hypotheses and evidence. Annu Rev Ecol Syst 11:197–232

Yeaton RI (1990) The structure and function of the Namib dune grasslands: species interactions. J Arid Environ 18:343–349

Young TP (1987) Increased thorn length in *Acacia depranolobium* – an induced response to browsing. Oecologia 71:436–438

Zak JC, Freckman DW (1991) Soil communities in deserts: microarthropods and nematodes. In: Polis G (ed) The ecology of desert communities. University of Arizona Press, Tucson, pp 55–88

Zemel A, Lubin Y (1995) Inter-group competition and stable group sizes. Anim Behav 50:485–488

Zumpt F (1965) Myiasis in man and animals in the Old World. A textbook for physicians, veterinarians and zoologists. Butterworths, London

Subject Index

Page numbers in italics refer to figures.